THE JOURNEYS OF
VOYAGER

This book was designed and produced by
Multimedia Books Limited
32-34 Gordon House Road
London NW5 1LP

Text copyright © Robin Kerrod 1990
Compilation copyright © Multimedia Books Ltd.

Editor: Anne Cope
Design: John Strange
Production: Hugh Allan

First published in the United States of America in 1990 by
The Mallard Press, an imprint of BDD Promotional Book Company, Inc.,
666 Fifth Avenue, New York, N.Y. 10103.

Mallard Press and its accompanying design and logo are trade
marks of BDD Promotional Book Company, Inc.

ISBN 0-792-45382-4

Typeset by O'Reilly Clark, London, England
Printed in Italy by Imago

THE JOURNEYS OF
VOYAGER

ROBIN KERROD

MALLARD
PRESS

Contents

6

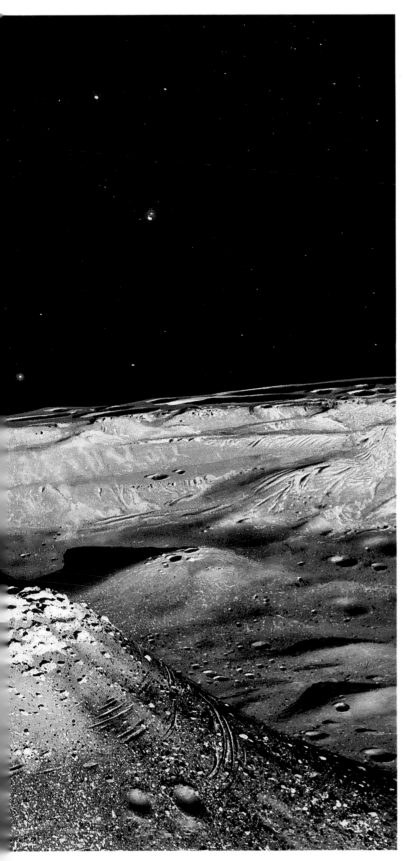

Journey of a lifetime

Billions of kilometers away in space, in the direction of the star constellation we know as the Great Dog, a spacecraft the size of a small car travels silently through space. From time to time, little hisses of gas shoot out from its thrusters, so that its dish antenna stays pointing in the same direction, towards what appears to be a bright star.

Periodically, bursts of radio waves emerge from the antenna and start traveling toward the bright star. But their destination is not the star, which we know better as the Sun, but a small body called the Earth. The Earth is but one of nine planets that form the major part of the Sun's family, or solar system. It takes more than four hours for the radio waves, even though they are traveling at the speed of light, to reach the Earth.

The name of this spacecraft is Voyager 2. It escaped from the clutches of Earth's gravity in 1977. It flew past the largest and most colorful planet in the solar system, Jupiter, two years later; in 1981 it sped past the beautiful ringed planet Saturn; in 1986 it visited the topsy-turvy planet Uranus; and in 1989 it approached the deep blue, cloud-flecked planet Neptune.

Neptune was its final port of call in the solar system, reached after a 12-year trek extending over 7 billion km (4.5 billion miles) of largely uncharted space. Far, far away, headed in another direction in space, is an identical craft, Voyager 1, which blazed a trail for Voyager 2 to Jupiter and Saturn.

Through the TV eyes of these space Voyagers we have seen enthralling sights. Raging storms on Jupiter, bigger than the Earth itself. Volcanoes on Jupiter's vivid moon Io, which spew out molten sulfur. Myriads of rings around Saturn, attended by tiny "shepherd" moons that keep the rings in place. Remarkable landscapes on Uranus's moon Miranda, which was once shattered to pieces in a catastrophic collision. Volcanoes of liquid gas on Neptune's largest moon Triton, the coldest place we know in the solar system.

All this and much more the Voyagers have revealed to us. For planetary scientists, for the whole human race, it has been the journey of a lifetime. No space probe will be able to repeat such a journey until the decade of the 2150s. Only then will celestial mechanics conspire once again to place the planets in a favorable alignment in the heavens.

By then the Voyagers will be out of contact with the Earth, but they will still be blazing trails — trails to the stars. And they are carrying to the stars and to any advanced civilizations they may encounter *en route* a message of good will from all mankind. A message in a bottle has been cast into the cosmic ocean. On what celestial shores will it come to rest?

◄The ringed planet Saturn, as it would appear from its small moon Mimas. This realistic painting was inspired by the images the Voyager probe returned on its epic journeys of discovery.

The Voyager Project

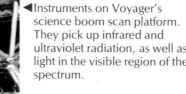
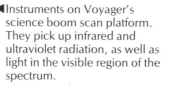

◀ Instruments on Voyager's science boom scan platform. They pick up infrared and ultraviolet radiation, as well as light in the visible region of the spectrum.

▶ (Opposite) The path of Mariner Jupiter Saturn through the solar system. Before the probe was launched, in summer 1977, it was renamed Voyager. The picture shows the probe in its final configuration.

In the early 1960s, when space scientists began to meet with some success in sending space probes to the near planets Venus and Mars, their thoughts inevitably turned to the giant outer planets. How could they explore these giants in a reasonable time frame, given that they lie hundreds of millions, even billions, of kilometers away?

Mars sometimes comes within 60 million km (37 million miles) of Earth. But Jupiter lies nearly three times as far away as Mars; Saturn, nearly twice as far as Jupiter; and Uranus, twice as far away again. Translate this into measurements and the magnitude of the problem becomes evident. Uranus, for example, is nearly 3 billion km (2 billion miles) away! Even using the most powerful rockets of the day, a flight to Uranus would have taken decades.

In 1965, however, a graduate student at Caltech (California Institute of Technology) named Gary Flandro came up with a brilliant notion: Why not use the gravity of one planet to accelerate a space probe and "sling" it toward another planet? The concept of the gravitational "slingshot", or gravity-assist method, was born.

"My gravity-boost idea wasn't new," recalled Flandro in 1989. "Astronomers had known for a long time that a comet speeds up when it passes close to a planet. I was the first to apply the same idea to a spaceship."

Gravity-assist meant that space scientists could give a space probe a much greater speed than was possible by means of a direct launch from Earth. It promised a much shorter journey time to the outer planets, provided they were favorably placed in the heavens.

Planet		Mercury	Venus	Earth	Mars	Jupiter	Saturn	Uranus	Neptune	Pluto
Diameter	(km)	4,878	12,104	12,756	6,794	142,800	120,000	50,800	49,100	2,280
at equator	(miles)	3,048	7,565	7,972	4,246	89,200	75,000	31,760	30,700	1,430
Mean distance	(km)	60	108	150	228	778	1,430	2,870	4,500	5,900
from Sun ($\times 10^6$)	(miles)	37.5	67.5	93.5	143	486	890	1,790	2,810	3,690
Circles Sun in		88 days	225 days	365.25 days	687 days	11.9 years	29.5 years	84 years	165 years	248 years
Spins on axis in		58.6 days	243 days	23.93 hours	24.6 hours	9.8 hours	10.2 hours	16.3 hours	16 hours	6.3 days
Mass (Earth=1)		0.06	0.82	1.00	0.11	318	95	14.5	17.2	0.002
Volume (Earth=1)		0.05	0.88	1.00	0.15	1316	755	52	44	0.005
Density (water=1)		5.4	5.2	5.5	3.9	1.3	0.7	1.3	1.8	2
Number of moons		0	0	1	2	16+	22+	15+	8	1

▼This diagram shows the orbits of the planets drawn roughly to scale. Note the enormous distances of the outer planets from the Sun, compared with the inner planets. All the planets travel around the Sun in an anticlockwise direction, as viewed from the "north" of the solar system. All the planets except one travel in more or less the same plane. The exception is Pluto, usually the outermost planet.

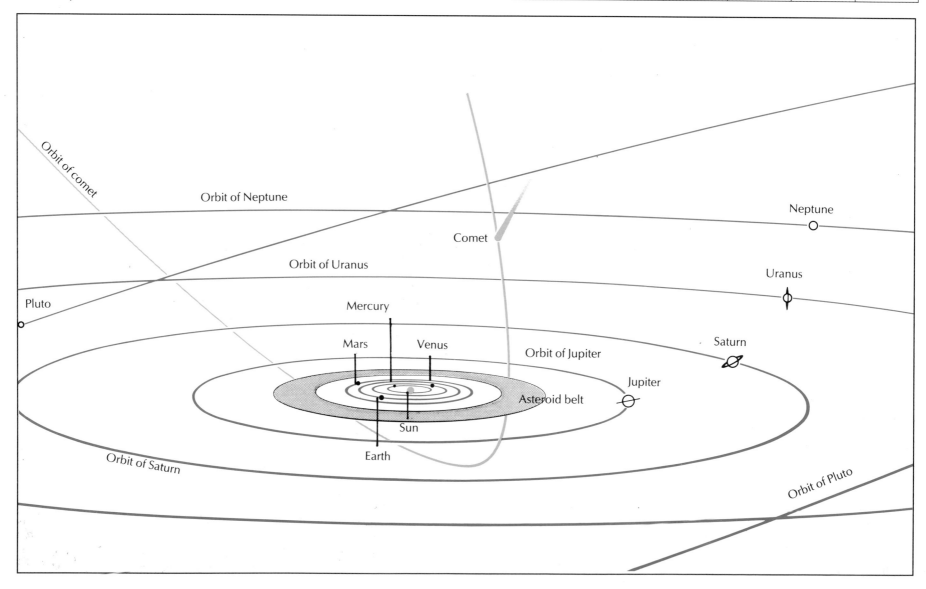

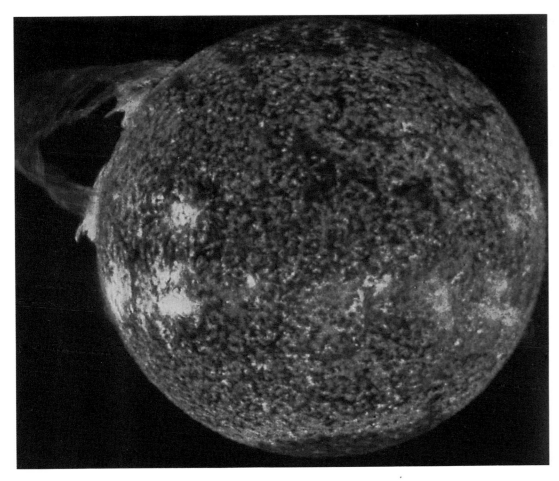

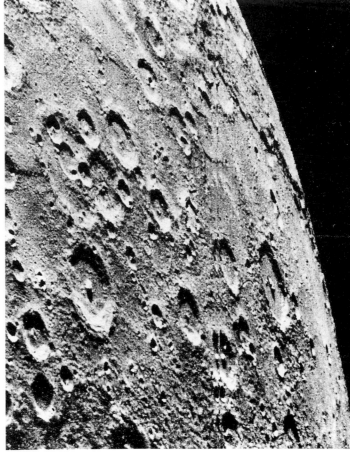

The Grand Tour

Fortune was truly smiling on planetary scientists at that time for, in the late 1970s and early 1980s, an alignment of the giant planets was imminent, an alignment that would not be repeated for another 176 years. A spacecraft launched in the late 1970s could use Jupiter's gravity to boost it on to Saturn, Saturn's gravity to boost it on to Uranus, and Uranus's gravity to boost it on to Neptune.

This multiplanet mission became known as the Grand Tour. It was a concept that captured and excited the imagination. Providing a suitable craft could be designed and deep-space tracking and communications problems could be mastered, only two things stood in the way of a successful Tour: the asteroid belt and Jupiter's powerful radiation. Both were capable of crippling a spacecraft, the one physically, the other electronically.

In the heady days of the Apollo Moon landings of 1969 the United States Congress approved a reconnaissance mission to Jupiter to investigate the possible asteroid and radiation hazards. The project, Pioneer Jupiter, was assigned to NASA's Ames Research Center at Mountain View, California. Two craft, Pioneer 10 and 11, were launched to Jupiter, one in 1972, the other in 1973. The 1973 launch of Pioneer 11 was into a trajectory that gave mission controllers the option of using gravity-assist to reach Saturn.

▲ (Left) The Sun is the only body in the solar system to emit light of its own. This is a Skylab image, which shows an enormous loop prominence. It is solar activity like this that makes the solar wind "blow" harder than usual.

▲ (Above) Mercury, the innermost planet, is a barren, Sun-baked world, which looks much like the Moon. It has no atmosphere and is peppered with craters. Like Earth, it is a rocky planet.

NASA began work on a Grand Tour strategy in 1969. It took note of a study by the National Academy of Sciences entitled "The Outer Solar System", which recommended that the United States undertake an in-depth planetary exploration program in the 1970s. Another Academy report two years later stated: "An extensive study of the outer solar system is recognized by us to be one of the major objectives of space science in this decade."

By then NASA had drawn up plans for a series of Grand Tour launches to Jupiter, Saturn and Pluto, beginning in 1976 and 1977, and to Jupiter, Uranus and Neptune, beginning in 1979. They had invited teams of scientists to come up with specific objectives for the missions. Caltech's Jet Propulsion Laboratory (JPL) at Pasadena, California, was carrying out design studies on an advanced spacecraft. The projected cost of the Grand Tours in the 1970s was about $750 million.

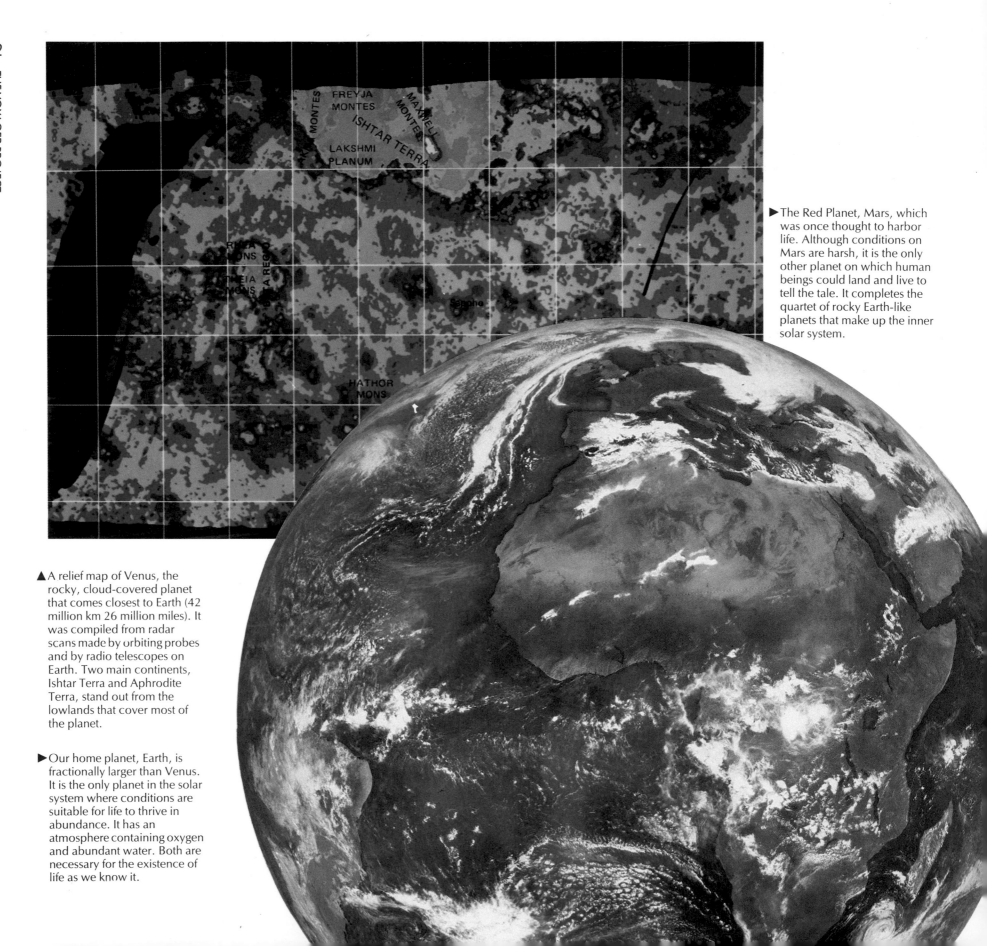

FREYJA MONTES

MAXWELL MONTES

ISHTAR TERRA

LAKSHMI PLANUM

HATHOR MONS

▶The Red Planet, Mars, which was once thought to harbor life. Although conditions on Mars are harsh, it is the only other planet on which human beings could land and live to tell the tale. It completes the quartet of rocky Earth-like planets that make up the inner solar system.

▲A relief map of Venus, the rocky, cloud-covered planet that comes closest to Earth (42 million km 26 million miles). It was compiled from radar scans made by orbiting probes and by radio telescopes on Earth. Two main continents, Ishtar Terra and Aphrodite Terra, stand out from the lowlands that cover most of the planet.

▶Our home planet, Earth, is fractionally larger than Venus. It is the only planet in the solar system where conditions are suitable for life to thrive in abundance. It has an atmosphere containing oxygen and abundant water. Both are necessary for the existence of life as we know it.

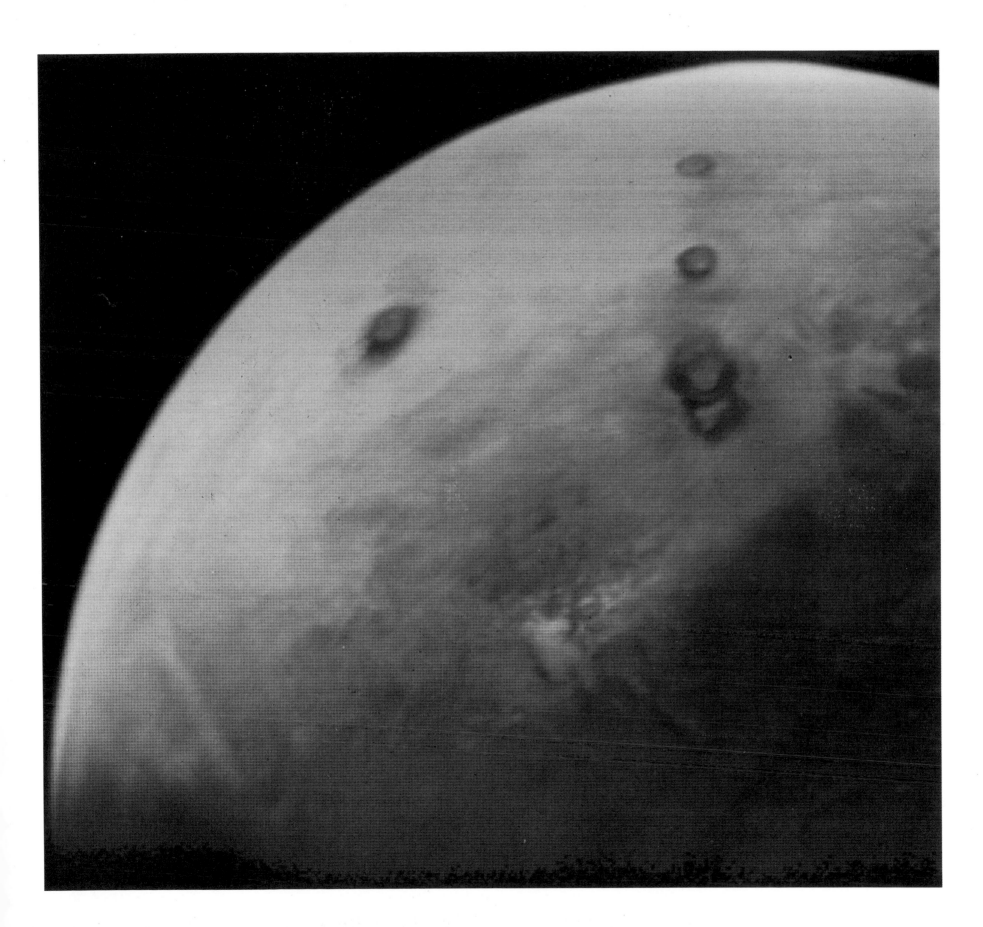

► The Jet Propulsion Laboratory (JPL) campus at Pasadena, close to Los Angeles in California. Part of Caltech, the California Institute of Technology, it manages the Voyager Project for NASA.

▼ In the Visitors Center at JPL stands a full-scale model of the first American satellite, Explorer 1, which JPL built and which was launched on 31 January 1958.

Mariner Jupiter Saturn

In 1972, however, budget cutbacks and a surprising disenchantment with space following the completion of the Apollo program put an end to the visionary concept of the Grand Tours. To salvage something from their grandiose plans, NASA put forward a new proposal for just two "no-frills" missions to Jupiter and Saturn, at one third of the original cost. It was named Mariner Jupiter Saturn, since the spacecraft would be based on an existing Mariner design.

The Mariner Jupiter Saturn project was approved and began officially on 1 July 1972. NASA invited detailed proposals for scientific objectives and instruments for the missions. Of 77 proposals received, 28 were selected. JPL was to be lead center for the project. By the year's end members of the science teams that would support the missions had been selected. Edward Stone of Caltech was appointed project scientist to head the science teams and coordinate all science activity.

The science teams, engineers, controllers, technicians and industrial contractors had less than five years to develop the necessary hardware and software. The deadline for launching the two mission spacecraft was the summer of 1977. By this time the project had been re-named Voyager.

Sophisticated robots

Two identical spacecraft, Voyager 1 and 2, were built to follow the trail to the outer planets blazed by Pioneer 10 and 11. They would be the most advanced craft that had ever been sent into space, sophisticated robots that to a large extent could act and think for themselves. They had to be self-sufficient because they would be traveling far, far away into a world of lengthy time-lags. No controllers on Earth could deal with real-time, on-the-spot emergencies that might happen billions of kilometers away. They would only learn about problems hours later, when signals, or the lack of them, would tell that something unexpected had occurred.

When in 1977 the Voyager craft were ready for lift-off, they were given a nominal design lifetime of five years — more than enough for them to reach Saturn.

"That didn't prevent us," recalled Edward Stone in 1989, "from launching Voyager 2 on exactly the same trajectory that the Grand Tour would have been on, hoping it would survive for 12 years."

Survive it did, and indeed still docs. So let us take a closer look at this indefatigable robot, which left the Earth atop a Titan-Centaur launch vehicle at 10:29 EDT (Eastern Daylight Time) on 20 August 1977.

The subcompact space traveler

Voyager has typical dimensions of about 3 meters (10 feet) across, and a mass of some 815 kg (1,800 pounds). In fact, it has approximately the same dimensions and weight as a subcompact car. Its most dominant feature is a parabolic reflector for the high-gain radio antenna. This reflector, which measures 3.7 meters (12 feet) across, is generally locked into a position facing Earth so that it can receive and transmit signals at maximum strength. It is kept in the right position by means of sensors that lock on to the Sun and Canopus, one of the brightest stars in the heavens.

Behind the reflector is a decagonal, or ten-sided, ring structure of aluminum, which houses ten electronics compartments. Tucked away in the middle of the decagon is a spherical tank made of titanium that contains hydrazine propellant for the spacecraft's propulsion system. This system feeds propellant to 16 little rockets, or thrusters, located at various points on the spacecraft.

Sets of attitude-control thrusters are fired automatically during flight to stabilize the spacecraft about its three axes. Another set of thrusters also fires on command to bring about a velocity change, which will alter the path, or trajectory, of the spacecraft. Such firings, called trajectory correction maneuvers (TCMs), are necessary to fine-time the trajectory of a spacecraft for a specific aimpoint during a planetary encounter.

Three main units project from the central structure of Voyager. One is the RTG boom. This carries three radioisotope thermoelectric generators (RTGs), which provide the electricity to power the spacecraft's instruments and electronics. They make electricity from the heat energy produced by the radioactive breakdown, or decay, of the plutonium isotope Pu-238.

RTGs have to be used to power deep-space probes. Solar cells, the standard power source for near-Earth spacecraft, cannot be used because there is not enough energy in the sunlight that filters through to the outer solar system.

At launch the total power available from the RTGs was more than 450 watts, but this has now fallen to less than 400 watts. The power output is continuous and cannot be switched off, so the excess is fed to a so-called shunt radiator, which radiates it into space as heat.

Located 180° from the RTG boom is the science boom, which carries most of Voyager's scientific instruments. Three of the instruments are fixed to the boom. The others are mounted on a platform that can move, or scan, on two axes. The scan platform allows precision pointing of these instruments, which include two TV cameras.

Two of the four magnetometers Voyager carries are mounted on the longest single structure of the spacecraft, the 13-meter (43-foot) long extendable boom. Two whip antennas 10 meters (33 feet) long complete Voyager's appendages.

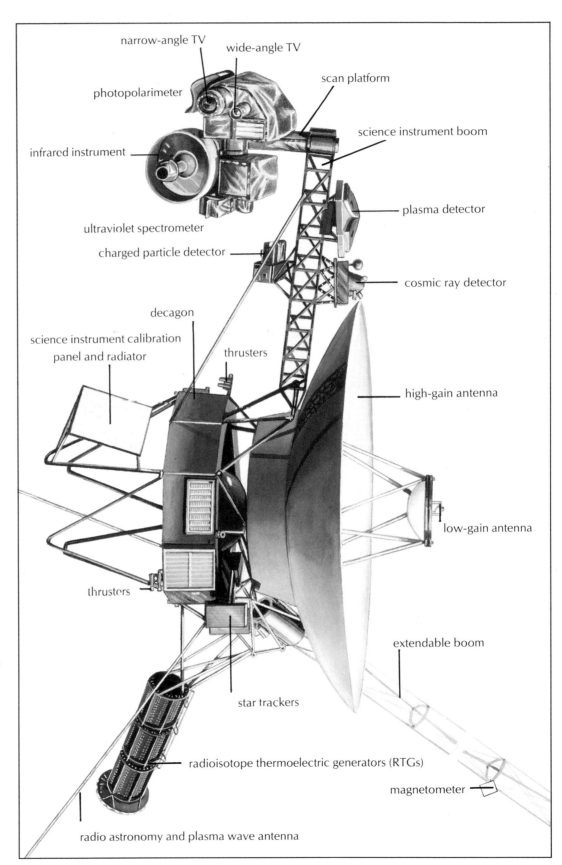

narrow-angle TV
wide-angle TV
photopolarimeter
scan platform
science instrument boom
infrared instrument
plasma detector
ultraviolet spectrometer
charged particle detector
cosmic ray detector
decagon
thrusters
science instrument calibration panel and radiator
high-gain antenna
low-gain antenna
thrusters
extendable boom
star trackers
radioisotope thermoelectric generators (RTGs)
magnetometer
radio astronomy and plasma wave antenna

▲ Voyager undergoes final testing before installation into the launch vehicle. The RTG boom (left) and science boom (right) are folded in their launch positions. The propulsion module, in place beneath the decagon, will provide the final boost to inject Voyager into the correct trajectory.

scene is then recreated, pixel by pixel and line by line.

Voyager's TV system produces a picture of 800 lines, compared with the 525 lines (in the United States) of the ordinary TV system. It also has 800 pixels per line. So the whole picture is made up of 800×800, or 640,000 pixels. Electronically, the brightness of each pixel is represented by a code of eight bits (binary digits). (This is the same kind of code that computers use, made up of 0s and 1s.) This means that a complete picture requires $8 \times 640,000$ or 5,120,000 bits.

The science instruments

As well as imaging, Voyager can carry out ten other scientific investigations with an interesting mix of instruments. Explains project scientist Edward Stone: "Because we didn't know exactly what we were going to see at Io [one of Jupiter's moons], or in fact anywhere else — Jupiter, Titan, the rings of Saturn — NASA chose a wide range of instruments so that we were not prejudging what it was we were going to discover."

The three fixed instruments mounted on the science boom are detectors for low-energy charged particles, cosmic rays and plasma. Cosmic rays are highly energetic charged particles (such as protons) that bombard our solar system from outer space. Plasma is a "cloud" of gases, whose atoms have been split up into a mixture of separate ions (charged particles) and electrons.

Streams of plasma come from the Sun in a flow we call the solar wind. Plasma is also present within the magnetosphere of planets, the "bubble" in which a planet's magnetic field can be felt. The nature and extent of the magnetic field is studied by means of Voyager's four magnetometers, two on the decagon and two at separate locations on the magnetometer boom.

In addition to the two TV cameras, the scan platform carries three other remote-sensing instruments, the photopolarimeter and the infrared and ultraviolet spectrometers. The photopolarimeter measures and analyzes the way its targets reflect light. It is used to determine the composition of planetary atmospheres, whose constituents affect light in a characteristic way. It is also used to investigate the surface of satellites and to measure the size of particles in planetary rings.

The infrared interferometer spectrometer and radiometer (IRIS) measures electromagetic radiation at wavelengths beyond the red end of the visible spectrum of light. Not only can it measure the heat given out by a planet and its temperature, it can also determine what gases are present in the atmosphere, each of which has a characteristic infrared "signature". The ultraviolet spectrometer, which "sees" in wavelengths beyond the violet end of the visible spectrum, can also study the composition, temperature and pressure of planetary atmospheres.

The two long whip antennas are the sensors for the planetary radio astronomy experiment. They are designed to pick up radio emissions from the planets and elsewhere in space. They are also shared by the plasma-wave instrument, another plasma investigation. This system can, for example, detect lightning flashes in an atmosphere.

The imaging system

The most important of Voyager's science systems for both scientist and layman alike is the imaging system. This gives Voyager its "eyes". It has two TV camera "eyes", which both use telephoto lenses. One gives a relatively wide field of view; the other, a much narrower one. By modern standards the vidicon tubes in the cameras are primitive. But they have managed to return pictures of the most remarkable quality, especially in view of the dim lighting conditions in which they operate.

The imaging system provides color images by means of a rotating color filter wheel on each camera. Pictures of the same area are taken in rapid succession through blue, green and orange filters. When the color signals are returned to Earth, they can be combined to give a simulated natural color image.

Each picture is produced as a normal TV picture is, by a line-scanning process. The scene viewed by the camera lens is split up by the scanning process into a number of lines, and the varying brightness of each part, or picture-element (pixel), along each line is converted into varying electronic signals. These signals are transmitted back to Earth, where they are converted back into a pattern of brightness. The original

◄A Titan-Centaur with Voyager 2 aboard blasts off the launch pad at Cape Canaveral's Complex 41 on 20 August 1977. This was the start of an extraordinary interplanetary journey that would eventually take in the four giant outer planets.

▼Voyager 1 casts a backward glance towards us on 18 September 1977 as it speeds into the depths of space, snapping a crescent Earth with a crescent Moon. It is now 12 million km (7.5 million miles) away.

▲ The 64-meter (210-foot) antenna at the DSN's Goldstone tracking station in the Mojave Desert in California. It is upgraded to 70 meters (230 feet) in time for Voyager 2's final encounter, with Neptune, in 1989. The Mojave site is perfect because the antenna is shielded from terrestrial radio interference by the surrounding hills.

▲ (Inset) Also for the Neptune encounter, the 27 dishes of the Very Large Array radio telescope at Socorro, in New Mexico, are integrated into the DSN communication network.

The main radio system for communications between the spacecraft and the Earth also dovetails as a radio science instrument. It is used to probe planetary atmospheres. When Voyager passes behind a planet, its radio waves pass through the atmosphere before they are extinguished. From the way they are affected, the density and other properties of the atmosphere can be determined. The same technique can be used to estimate the size of particles in a planet's rings.

The brains of Voyager
Because Voyager was designed to travel very far from Earth, it had to be equipped with sufficient "brainpower" to enable it to operate, if necessary, for long periods of time without human intervention. Accordingly, it was built around three main computers, each of which had a redundant, or backup, system.

The top "brain" is the computer command system (CCS), which coordinates all on-board activities using instructions stored in its memory. These instructions can be altered by painstaking reprogramming from Earth. The CCS computer is hardly "state of the art". Its random access memory (RAM) has only about one hundredth of the capacity of today's personal computers!

The CCS also controls the other two computers, the flight data system (FDS) and the attitude and articulation control subsystem (AACS). The FDS controls all the science instruments and encodes all science and engineering data for transmission to Earth. The data can be transmitted in real time, that is, as soon as it is acquired, or it can be stored on a digital tape recorder for delayed transmission. The AACS controls Voyager's position, or attitude, in space. It operates the thrusters for spacecraft maneuvering and moves the scan platform.

The Deep Space Network
Voyager has been controlled from the outset by JPL (Jet Propulsion Laboratory) at Pasadena. JPL is responsible for tracking and commanding the spacecraft, and handling and processing the data it transmits. The key element in spacecraft communications is NASA's Deep Space Network (DSN) of tracking stations, which is also operated by JPL.

Three DSN stations are positioned at widely separated longitudes around the world, so that at least one is always "in sight" of Voyager as the Earth rotates. One is located at Goldstone, in California's Mojave Desert; a second at Robledo, near Madrid, Spain; and a third on the Tidbinbilla Nature Preserve, near Canberra, Australia. Each station is now equipped with antennas 70, 34 and 26 meters (230, 112 and 85 feet) in diameter.

The telemetry, or transmitted data, received at Goldstone is sent directly to JPL through a microwave link. Telemetry received at the overseas stations is routed to JPL via satellite links through NASA's Goddard Space Flight Center at Greenbelt, Maryland.

At JPL the data are fed to the Network Operations Center, where they are logged on tape and then routed to the Mission Control and Computing Center (MCCC). The MCCC in turn sends data to the Test and Telemetry System, which displays science and engineering data in real time, and to the Multimission Image Processing Laboratory (MIPL) for processing.

In the MIPL the imaging data are decoded and converted into images. These images may be displayed on the monitor at Mission Control or processed into photographs. During the computerized conversion process, the images can be enhanced, that is, corrected and made more meaningful, by the use of false colors.

Space, time and distance
The farther Voyager travels away from Earth, the more difficult it becomes to maintain communications with it. The network of DSN stations tracks Voyager carefully so that their antennas always point precisely towards it when signals are being transmitted or received.

There is not too much of a problem with the uplink, or transmission from Earth to the spacecraft. The 70-meter (230-foot) diameter antennas can be used at 400 kilowatts power. The real worry is with the downlink, or transmission *from* the spacecraft. Voyager's radio operates at a power of only 20 watts, about the same as the light bulb in a refrigerator, and it has only a small dish antenna. By the time the radio signals reach Earth, they have billions of times less power than the battery that runs a digital watch! Nevertheless, the huge antennas of the DSN, electronically linked, can pick up the feeble signals, which JPL's computers can then convert into remarkable images.

The great distances between Earth and Voyager create another communications problem, the time-lag. Radio waves sent to the craft take a measurable amount of time to make the journey.

Like all forms of electromagnetic radiation, radio waves travel at the speed of light, some 300,000 km per second, or more than 1 billion km/h (186,000 miles per second, 670 million mph). On Earth distances are so small, that we think of light and radio waves as traveling instantaneously. But in space we must allow for a finite time of travel.

When Voyager 2 encountered Jupiter, it took some 47 minutes for its signals to reach Earth. By the time it reached Saturn, the journey time had increased to 1 hour 26 minutes; at Uranus, it was 2 hours 45 minutes; and at Neptune, no less than 4 hours 6 minutes! So the minimum time it took controllers to carry on a "conversation" with Voyager at Neptune was 8 hours 12 minutes.

As far as the Voyager mission is concerned there are two kinds of time, spacecraft time and Earth time. The two differ by the one-way time-lag according to how far away Voyager is. In the account of the Voyager missions that follows, it is specified whether the times of mission events are spacecraft times or not.

There is also, of course, a problem with Earth time, because of the different world time zones! The procedure adopted here is to quote mission event times in terms of JPL local time, that is Pacific Standard Time (PST) or Pacific Daylight Time (PDT). The only exceptions are the times of the Voyager launches, which are quoted in Eastern Daylight Time (EDT), because the launches took place at Cape Canaveral on the east coast of the United States.

Robot ambassadors
Sometime next century the Voyager probes, like the Pioneer craft that preceded them, will escape from the region of space dominated by the Sun and head into the interstellar void towards other star systems.

As robot ambassadors of the human race, it is appropriate that they carry with them an artifact that encapsulates life on this planet. It is a phonograph record called "Sounds of Earth". It carries messages and coded images that, one day in the remote future, may be deciphered by one of the other intelligent life forms that must exist somewhere in our galaxy (see page 76).

▼This is one of the best color photographs taken from Earth of the disc of Jupiter. Astronomers at Kitt Peak National Observatory in Arizona took it with a 2.1-meter (84-inch) telescope.

2 Encounter with Jupiter

In Roman mythology Jupiter was king of the gods and ruler of the heavens. It is an appropriate name for the largest planet in the solar system — Jupiter is more than twice as massive as all the other planets put together. Its diameter is 11 times that of Earth.

In several respects Jupiter resembles a star rather than a planet such as Earth. Whereas the Earth is solid, Jupiter is fluid. Its density is low and comparable with that of our nearest star, the Sun. It is composed mainly of hydrogen and helium. It has a powerful magnetic field, radiates heat, and gives off radio waves and even X-rays. There is little doubt that, if Jupiter had been very much bigger, it would have begun to shine as a star, and our solar system would have become a binary, or double-star, system.

Of course, Jupiter does shine, but by reflected sunlight. Even though it never comes closer to us than about 600 million km (370 million miles), it outshines all the other planets except Venus almost all of the time. Being so conspicuous, Jupiter was familiar to ancient astronomers. The Italian astronomer Galileo first trained a telescope on the planet in 1610 and was astounded to see four moons circling around it. This discovery was a milestone in astronomy for it provided evidence in favor of a Sun-centered rather than an Earth-centered universe.

Through the telescope
Even through binoculars we can easily see Jupiter as a bright disc. We can also see the four moons spotted by Galileo as bright points on either side. They are, in order of distance from the planet, Io, Europa, Ganymede and Callisto. Ganymede is the largest moon in the solar system, similar in size to the planet Mercury.

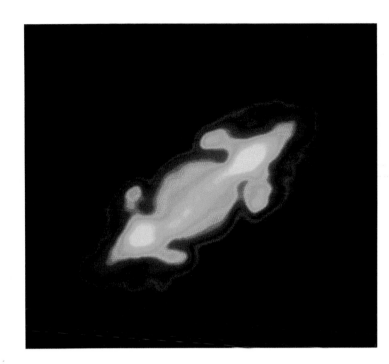

The color of the Spot varies from orange and pink to brick-red. The Spot always remains at the same latitude (about 20° south), although it tends to drift in longitude. It measures about 14,000 km (9,000 miles) wide and some 28,000 km (17,500 miles) long. It was once thought to be a projection from the planet's surface but now appears to be a massive storm center. Other persistent features, such as the system of white ovals south of the Spot, also appear to be raging storms.

The Pioneer pathfinders

The probes that preceded the Voyagers to Jupiter could not have been better named, for they were indeed pioneers of the deep space frontier, blazing a trail to the outer planets.

Pioneer 10 left Earth in March 1972 at more than 51,500 km/h (32,000 mph), the highest speed yet reached by any man-made object. It was headed for a close encounter with Jupiter 21 months later. In prospect was a journey of nearly 1 billion km (625 million miles). It was a potentially hazardous one because, to get to Jupiter, Pioneer would have to traverse

◄Another way of "seeing" Jupiter from Earth is at radio wavelengths, since the planet is a strong radio source. This is the kind of image obtained by processing Jupiter's radio signals.

▼Some of the far-flung tiny moons of Jupiter are undoubtedly captured asteroids, and may well look like this. This is in fact Phobos, one of the two moons of Mars, but it is probably also a captured asteroid.

Were it not for the brilliance of their parent planet, the Galilean moons would be visible to the naked eye. They appear to move back and forth in a straight line because they orbit in the plane of Jupiter equator, which we see edge-on.

Another nine smaller moons can be seen in a powerful telescope. One, Amalthea, lies closer in than Io, but the others orbit the planet very much farther out, in two widely separated groups of four. The most distant group of moons circles Jupiter clockwise (from east to west). This is a retrograde motion — in the opposite direction from most of the other bodies in the solar system. It is thought that they are captured asteroids.

Belts, zones and spots

When viewed through a telescope, Jupiter is a riveting spectacle. The most noticeable feature on the disc is the system of roughly parallel dark and light bands, which are known respectively as belts and zones. Closer inspection of the disc reveals a wealth of finer detail — features variously described as ovals, spots, festoons, wisps, streaks, condensations and plumes. Prolonged observation of these features shows that the planet, or rather its gaseous surface, is rotating rapidly.

The rapid rotation period, less than 10 hours, is the fastest of any planet. It causes the gaseous ball that is Jupiter to bulge at the equator and flatten at the poles. This flattening can be seen quite clearly in a telescope. The rapid rotation is also responsible for the permanent system of belts and zones. These are clouds that have been stretched into bands parallel with the equator by the force of rotation.

The fascinating variety of features that appear in the belts and zones are disturbances within the general circulation pattern. The biggest and most persistent of them is called the Great Red Spot, first recorded more than three centuries ago.

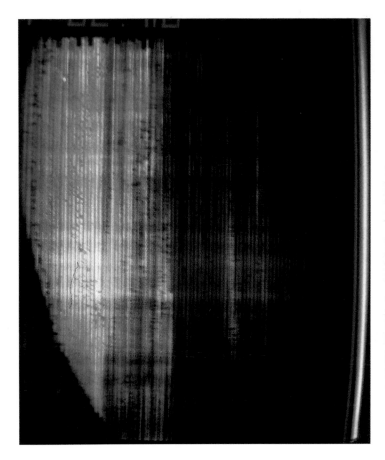

◀(Opposite) Pioneer encounters the king of the planets, Jupiter, in a painting executed before the Pioneer probes set off on their pathfinding missions. Note the differences between Pioneer and Voyager.

◀(Left) Raw imagery recorded at Ames Research Center from signals sent back by Pioneer 10 about two hours before closest approach to Jupiter on 3 December 1973. Later it will be computer-enhanced and corrected to provide the best pictures of Jupiter yet.

◀James Van Allen, principal investigator for Pioneer's charged particles experiment, is on hand at Ames for the first Pioneer Jupiter encounter. It was Van Allen who discovered the "belts" of radiation surrounding the Earth.

the asteroid belt, the broad band of mini-planets and interplanetary debris that exists between the orbits of Mars and Jupiter. But, much to the relief of the mission scientists at Ames Research Center, the spacecraft emerged unscathed from the belt.

As Pioneer drew near to Jupiter, it began returning images much better than any taken through Earth telescopes, showing the planet's turbulent and multicolored atmosphere and the enigmatic Great Red Spot. It made its flyby of Jupiter on 3 December 1973, passing 131,000 km (83,000 miles) above the cloud tops. Its instruments suffered some damage during passage through Jupiter's radiation belts, which proved to be much more intense than expected. After close encounter, Jovian gravity grabbed Pioneer and flung it out into the void of space at sufficient velocity to enable it to escape, eventually, from the solar system.

At a press conference at Ames, NASA planetary program director Robert Kraemer enthused about the mission's success: "We sent Pioneer off to tweak a dragon's tail, and it did that and more. It gave it a really good yank, and it managed to survive."

On 13 June 1983 Pioneer 10 crossed the orbit of Neptune, currently the farthest planet from the Sun, and began traveling through interstellar space. This event, observed astronomer Carl Sagan, was "filled with symbolism. Lots of things have entered the solar system during its 5-billion-year history — comets, asteroids, every sort of cosmic debris. This

is the first time, however, that something associated with life and intelligence has left it."

Pioneer 11: the second Jupiter encounter
The launch of Pioneer 10's near-identical sister craft, Pioneer 11, was deliberately delayed so that mission scientists could benefit from Pioneer 10's experiences. The basic flight plan was chosen so that Pioneer could not only encounter Jupiter but, all being well, press on to fly by Saturn.

Pioneer 11 lifted off in April 1973 and, like its sister craft, emerged from the asteroid belt unscathed. This confirmed the good news for planetary scientists that the asteroid belt is not, as had been feared, a formidable barrier for deep-space exploration.

Pioneer looped behind Jupiter on 2 December 1974 and flew within 43,000 km (27,000 miles) of the Jovian cloud tops. After the encounter Pioneer began its near five-year long cruise towards Saturn, aiming for a close approach in September 1979.

Pioneering results
Here are some of the findings made by the Pioneers. Jupiter is almost entirely made up of hydrogen, together with a little helium. It is probably a totally liquid planet, without any solid surface at all. The colorful dark belts and light zones are clouds stretched out by the rapid rotation of Jupiter's atmosphere. The belts appear to be regions of cool, dry

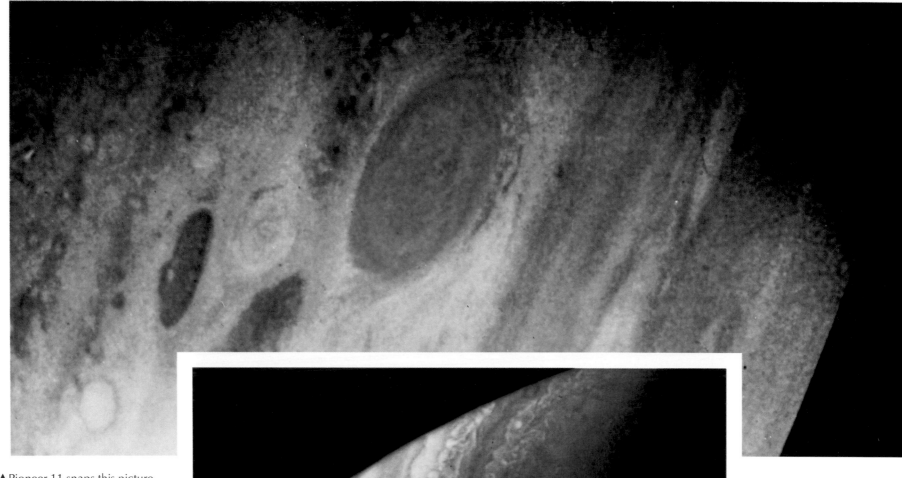

▲ Pioneer 11 snaps this picture of Jupiter's Great Red Spot from a distance of 545,000 km (338,000 miles). It shows the complex flow patterns in the surrounding atmosphere, and hints at circulation within the Spot itself.

► Part of the north polar region of Jupiter, pictured by Pioneer 11 in blue light. There is violent turbulence in the atmosphere.

descending gas; the zones, regions of warm, moist ascending gas. Much of the heat that powers the weather system comes from inside the planet. This internal heat source makes Jupiter radiate into space nearly twice as much heat as it receives from the Sun.

Jupiter's magnetic field is about 2,000 times stronger than the Earth's, and its radiation belts are up to 10,000 times more intense. The magnetosphere is a source of high-energy electrons, which have sometimes been detected on Earth. The giant planet is also a strong radio source, second in power in the heavens only to the Sun. Enormous radio blasts are generated when electricity flows between Jupiter and the inner Galilean moon Io along magnetic field lines.

Voyager 1: the mission begins

Voyager 1 began its journey to Jupiter 16 days after its sister craft, on 5 September 1977. But it was launched on a shorter, faster trajectory so that it would be the first to reach Jupiter and afterwards Saturn. On 10 December Voyager 1 entered the asteroid belt neck and neck with Voyager 2, but five days later began to pull away. It would travel inside the belt for another nine months.

As far as particle hits were concerned, it was to be an uneventful crossing. But mission scientists had other worries. On 23 February 1978 Voyager's scan platform jammed during a routine calibration maneuver. If this could not be freed, it would be disastrous. The platform carried the TV cameras and other major instruments, which had to be aimed with high precision to acquire the desired targets.

Throughout March and April tests were carried out to move the scan platform slowly, a little at a time. Not until May were mission engineers confident enough to put the scan platform through its proper paces. Fortunately, it behaved normally. They suspected that the problem must have been precipitated by a small piece of soft debris catching in the gears. Gradual maneuvering of the platform must have either crushed it or pushed it aside.

A different Jupiter

During the remaining months of 1978 Voyager returned progressively better images of Jupiter. Its other instruments probed the interplanetary medium in general and the region of space around the giant planet in particular. By 10 December Voyager was returning higher-resolution images than could be obtained from telescopes on Earth.

On 4 January 1979 Voyager was 60 million km (38 million miles) from Jupiter and almost exactly two months away from closest approach. The so-called observatory phase of the mission began, when all science activity was directed towards the planet, its moons and the surrounding environment. Voyager began taking images of Jupiter through colored filters to produce simulated natural color pictures. It soon became evident that the appearance of Jupiter's banded atmosphere had changed markedly since 1974.

As February progressed, mission scientists at JPL became progressively more excited. Imaging team scientist Garry Hunt, seconded from University College, London, said of

◀On 5 September 1977 Voyager 1 blasts off from Cape Canaveral, more than two weeks behind its sister craft. But in a little over three months, it will nose in front, and in March 1979 reach Jupiter first.

▶Jupiter looms large and clear in Voyager 1's cameras on 5 February 1979. The picture shows not only the famous Great Red Spot, but also the colorful moon Io making a transit across the disc. Other circular storm centers can be seen, particularly in the more northerly and southerly regions.

Jupiter: "It seems to be far more photogenic now than it did during the Pioneer encounters. I'm more than delighted by it … There are infinitely more details than we ever imagined."

The sulfur mystery

The near-encounter period of the Jupiter flyby opened officially on 22 February 1979, with a press conference at NASA Headquarters in Washington DC. Some of the results of the observatory and far-encounter phases were presented. Auroral activity had been observed in Jupiter's atmosphere, and new low-frequency radio emissions had been detected. Ultraviolet radiation was coming from a doughnut-shaped ring, or torus around the planet at the orbit of Io. This was quite baffling.

On 27 February the ultraviolet investigators suggested that the ultraviolet radiation could be produced by a plasma of sulfur ions — sulfur atoms that had lost electrons. But what sulfur ions were doing in the region was anybody's guess. Also that day mission scientists were expecting Voyager to meet Jupiter's bow shock. The Pioneer spacecraft had met it about 7 million km (4.5 million miles) out from the planet. But Voyager did not detect the bow shock until early next morning

▲ From a distance of about 20 million km (12.5 million miles), a wealth of fascinating features begin to come into focus in the stormy Jovian atmosphere. Io is again visible, this time above the Great Red Spot. At the right is another moon, Europa, seemingly close, but in reality several hundred kilometers away.

◄ From a distance of 8 million km (5 million miles) Jupiter's moon Ganymede displays some interesting markings, not yet explainable. But later views will reveal all. With a diameter of 5,276 km (3,278 miles), Ganymede is the largest moon in the solar system.

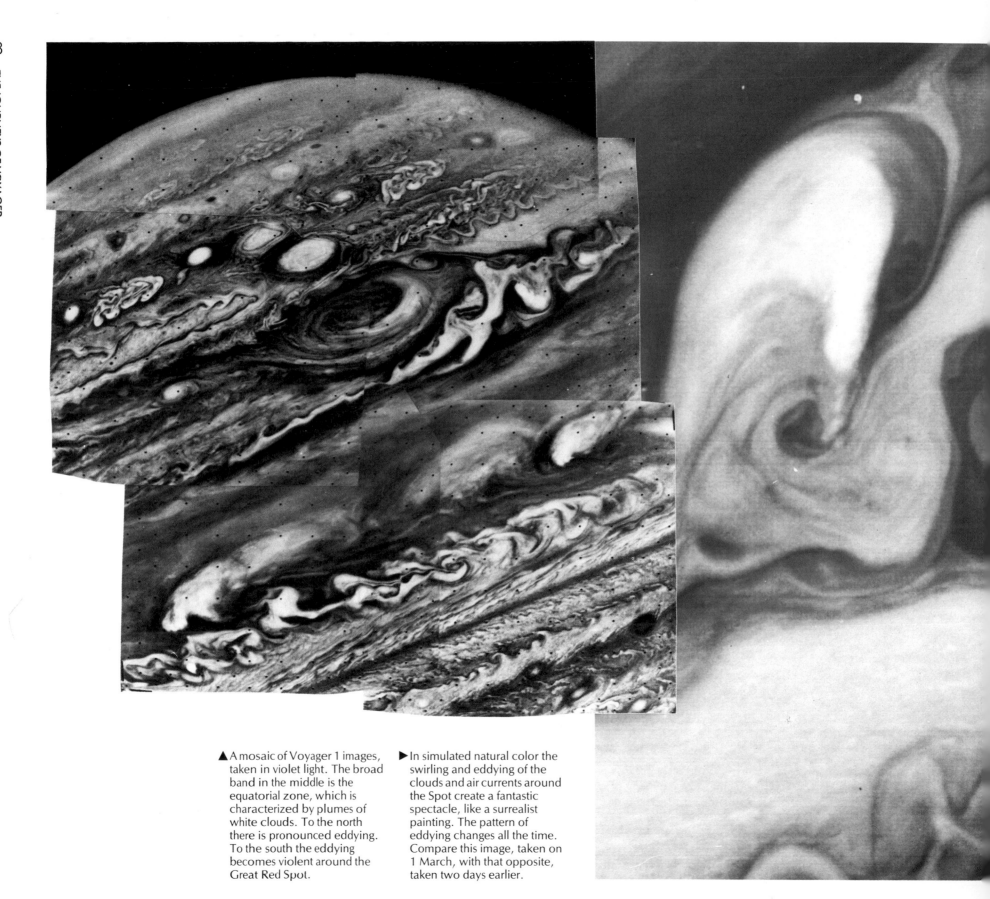

▲ A mosaic of Voyager 1 images, taken in violet light. The broad band in the middle is the equatorial zone, which is characterized by plumes of white clouds. To the north there is pronounced eddying. To the south the eddying becomes violent around the Great Red Spot.

▶ In simulated natural color the swirling and eddying of the clouds and air currents around the Spot create a fantastic spectacle, like a surrealist painting. The pattern of eddying changes all the time. Compare this image, taken on 1 March, with that opposite, taken two days earlier.

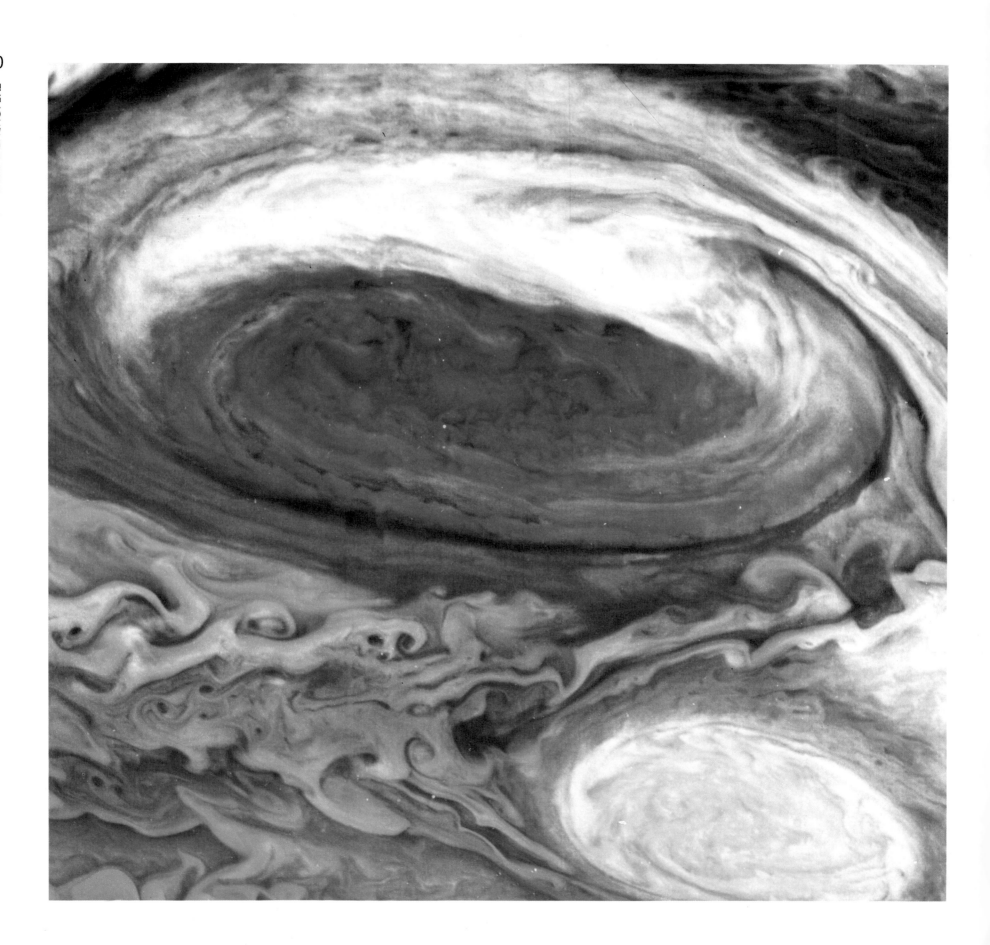

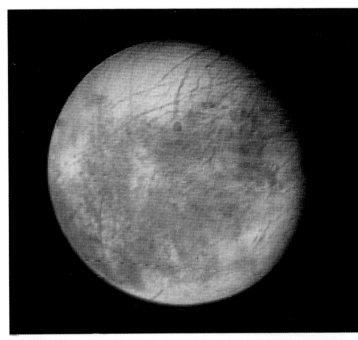

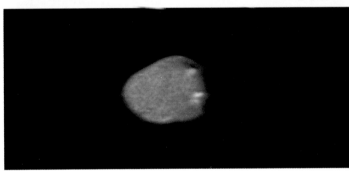

when it was nearly a million km (625,000 miles) closer in. The increased activity observed on the Sun since 1974 must have increased the strength of the solar wind, which must have compressed Jupiter's magnetosphere.

Bothered and bewildered

At 11:00 A.M. PST on 28 February, imaging team leader Bradford Smith opened the first daily press conference at the Von Karman Auditorium at JPL, as the world's media gathered for the upcoming close encounter. Confessed Smith: "After nearly two months of atmospheric imaging and perhaps a week or two of satellite viewing, we're happily bewildered."

Mission scientists were also bewildered to see Voyager cross the bow shock another four times over the next two days as the magnetosphere periodically expanded and contracted. They were becoming worried that the compression of the magnetosphere by the strong solar wind would lead to more intense radiation in Jupiter's radiation belts. If so, this might adversely affect Voyager's electronics and put the mission in jeopardy.

At the morning press conference on 3 March the latest

results and images were shown, as were time-lapse movies of the circulation in Jupiter's atmosphere. Close-up views showed the serpentine motions of the clouds around the Great Red Spot. Just south of the Spot three white ovals stood out. They were interesting historically, because astronomers had witnessed their birth 40 years before.

The "big four" Galilean moons of Jupiter were also beginning to take on distinctly different characters, which whetted the appetite of the satellite specialists. Io in particular appeared very strange, displaying circular, crater-like features and a curious "hoof-print" marking. By midday PST Voyager, now curving in towards the giant planet, had passed the orbit of Ganymede. It began sending back improved images of the moons: icy, striped Europa, tiny Amalthea and fascinating Io, which continued to intrigue. Here was no ordinary drab rocky moon, but one with a unique surface colored a vivid red, yellow and orange.

Five million amps

As Voyager raced towards its rendezvous with Jupiter, it was performing beyond expectation and sending back images of the kind that planetary geologists dream of. The only problems being experienced did not originate with Voyager, but back on Earth. Stormy weather in Australia had caused the loss of several hours of precious data at the DSN's Canberra antenna; problems at the DSN's Madrid tracking station had also caused data loss.

But mission scientists had more than enough to be going on with. Even the imaging team, said Bradford Smith, were "just standing around with their mouths hanging open watching the pictures come in." And the pictures were coming in at the rate of one every 48 seconds!

Monday 5 March was close-encounter day. In Earth skies in the early morning the "star of the show", giant Jupiter, shone brightly. Over 679 million km (422 million miles) away, at

◄(Opposite) The Great Red Spot, seen in close-up and in false color. The clouds within the Spot rotate anticlockwise about once every six days. They travel at a speed of about 100 meters (330 feet) per second.

◄(Top left) The day before encounter, Voyager 1 snaps this picture of Europa (diameter 3,126 km/1,942 miles), smallest of the Galilean moons. It appears to be covered with ice and is very bright.

◄(Bottom left) Tiny Amalthea is shaped rather like a potato, and is no more than 270 km (170 miles) across. It orbits the planet about every 12 hours. It was first discovered nearly a century ago, in 1892.

▼The "pizza moon" Io has a landscape unlike that of any other body yet encountered in the solar system, beautiful, bizarre and utterly baffling. Io has a diameter of 3,632 km (2,257 miles).

▶A few hours after closest approach to Jupiter, Voyager 1 swoops close to Io and transmits another remarkable picture. The "hoof print" feature is clearly seen here.

▼Computer enhancement of Io pictures produces an astonishing revelation: the moon is volcanically active. This picture shows an eruption occurring on the limb. Sulfur and dust are being blasted to heights of 200 km (125 miles) or more.

4:05 A.M. PST spacecraft time, Voyager made its closest approach to the planet, 278,000 km (173,000 miles) above the cloud tops. Some three hours later Voyager skirted the so-called flux tube, the region where scientists suspected huge electric currents (later estimated to be 5 million amps) were surging back and forth between Jupiter and Io. A few minutes later it made a close flyby of Io, before disappearing into Jupiter's shadow. There, out of radio contact with Earth, its instruments scanned the Jovian atmosphere for signs of auroral and lightning activity.

A pizza moon

Out into the sunlight once again, Voyager began playing back the data it had recorded during its two-hour blackout. Again it was Io that riveted the attention and stretched the imagination. Irreverently, Bradford Smith said that "Io looks better than a lot of pizzas I've seen!" And it has since often been called the pizza moon.

Apart from its striking colors, the strange thing about Io was the absence of impact craters. This meant that the surface was somehow being eroded or being modified by internal processes, and could not be more than 100 million years old. Or did Jupiter in some way protect its moons from bombardment from outer space? Were all Jupiter's large moons craterless?

Mission scientists soon got their answer. On the evening of encounter day pictures came in of Ganymede, largest moon in the solar system, and showed that its icy surface was peppered with impact craters. Io was clearly a special case, but what was it that had shaped or was shaping its surface?

Meanwhile it was becoming evident that Voyager had sustained some electronic damage on its passage through Jupiter's radiation belts. The main clock on the spacecraft, which timed every operation, was now running at least 6 seconds slow. In addition, two of the on-board computers had got out of synchronization with each other, and this had resulted in blurring of some of the high-resolution images of Io and Ganymede. The clock and computers were subsequently reset without any problems. All of the instruments survived the high radiation dose except the photopolarimeter, which failed a few hours before closest approach.

By Jove, a ring!

The days following encounter still brought forth surprises. The flyby of Callisto revealed a heavily cratered surface with a bullseye pattern of low ridges, which must have resulted from a massive impact. Images acquired in Jupiter's shadow revealed vigorous displays of auroras and "super bolts" of lightning.

An offchance camera shot taken as Voyager passed through the equatorial plane made a startling find that surprised even those who had become used to expecting the unexpected. It was announced by Bradford Smith at the 7 March press briefing. "This morning," said Smith, "I would like to add yet another important discovery to be claimed by this outstanding mission — that of a thin, flat ring of particles surrounding Jupiter. Thus Jupiter joins Saturn and Uranus as the third planet of our solar system known to possess a planetary ring system." The single ring appeared to extend out to a distance of about 57,000 km (36,000 miles) above the cloud tops.

At the final press briefing next day mission scientists gave an overview of Voyager's discoveries. Bizarre Io still defied explanation. But mission scientist Laurence Soderblom was unknowingly prophetic when he described the moon as having a surface "much like you might see around a fumarole at Yellowstone National Park." A fumarole is a volcanic vent.

The plumes of Io

Soderblom's remarks were prophetic because later that day mission navigation specialist Linda Morabito was checking an image of Io against a background of stars, snapped by Voyager early that morning. She saw what appeared to be an umbrella-shaped cloud on the sunlit limb of the moon. But Io did not have an atmosphere and so could not have clouds. What could this feature be? The only explanation appeared to be a violently erupting volcano — violent because the cloud material was being shot 300 km (200 miles) above the surface!

When other Io images were computer-enhanced, further volcanic clouds, or "plumes", were revealed. Io *was* volcanically active, the only other body in the solar system to be so besides Earth. Independently, the infrared investigators

had come to the same conclusion, having detected heat and sulfur dioxide gas near the plumes.

Here at last was an explanation for the sulfur ions found around Io's orbit. The presence of sulfur on the surface would also explain Io's vivid coloring. The plume that launched the volcano search came from a vent in the middle of the "hoof-print" feature noted earlier. This feature was named Pele, after the Hawaiian fire goddess. Other major volcanic features were called Prometheus, Loki and Marduk.

The slingshot effect of Jupiter's gravity had meanwhile accelerated Voyager to a speed of 135,000 km/h (84,500 mph), and its 750-million-km (470-million-mile) journey to the outermost planet known to the ancients, Saturn, had begun.

On 13 April Voyager's seven-month long surveillance of Jupiter, which had yielded more than 18,000 images, came to an end, and most of its instruments were put on standby. Mission scientists were now getting ready for the next Voyager spectacular, featuring Voyager 2, whose observatory phase was due to begin in less than two weeks.

▼ Jupiter's second largest moon, Callisto, has a surface of dirty ice. The bright spots are places where fresh ice has been exposed by meteorite impacts. The picture shows a huge impact basin named Valhalla — the central light area is about 600 km (375 miles) across. Callisto has a diameter of 4,820 km (2,995 miles).

▶ In July 1979 Voyager 2 spots changes in the circulation of the atmosphere around the Great Red Spot since the Voyager 1 flyby. For example, there is now a different white oval to the south. Compare this with the image on page 29.

▼ Jupiter's equatorial zone runs across the middle of this false-color Voyager 2 picture. Notice the typical plumes of white cloud. To the north is a dark brown oval, which appears to be a region where the upper ammonia cloud layer has parted to reveal darker clouds beneath. Brown ovals occur frequently in the northern hemisphere.

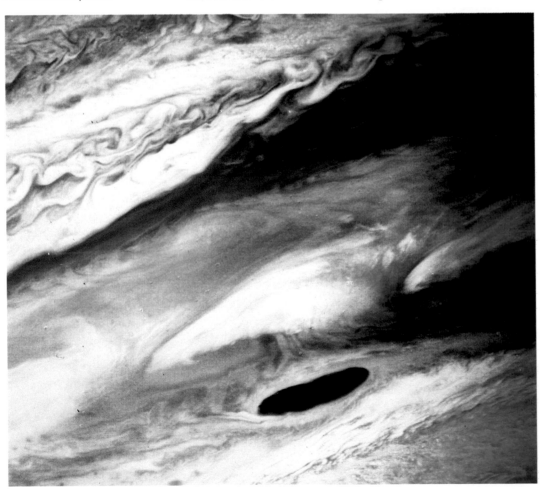

Voyager 2 at Jupiter

Voyager 2 had been launched on what was to prove the interplanetary mission of the century before Voyager 1, on 20 August 1977. It was sent on a slower trajectory than Voyager 1, which eventually overtook it on 15 December while both spacecraft were beginning their traverse of the asteroid belt.

Like its sister probe before it, Voyager 2 ran into a major problem in the spring of 1978, while still inside the asteroid belt. On 5 April its primary radio receiver suddenly failed. Mission engineers were not initially too worried, because the Voyagers have a backup receiver. However, when the backup was automatically switched in, they found that this was also faulty and they could not communicate with it! Here was a real crisis.

After 12 hours of non-communication, Voyager's command system automatically switched back to the primary receiver, as it had been programmed to do. This time the primary receiver worked, and engineers began sending instructions to it. But after half an hour a power surge in the circuits blew the fuses. The primary receiver died — this time for good. Because of the way it was programmed, Voyager's command system would not try to switch on the primary receiver again for seven days. Then, because this receiver was dead, it would switch in the backup.

Doppler effects and defects

The good news was that JPL engineers had by now realized what the fault was in the backup. Because of the failure of one of its circuits, it could only accept radio signals of a certain frequency. A healthy receiver could accept signals over a wide frequency range.

This facility is needed in space communications systems because of the Doppler effect, the change in frequency that occurs in radio waves when there is relative movement between transmitter and receiver. (The well-known change in pitch of a police-car siren as the car first approaches, then recedes from you, is the result of the Doppler effect in sound waves.)

The Doppler effect in space-probe communications is caused by the rotation of the Earth, which makes the transmitting antenna move alternately toward and away from the spacecraft. There is an additional Doppler effect during close-encounter periods when the spacecraft accelerates under gravity towards its target.

To counter the Doppler effects, the JPL engineers devised a new technique for transmitting radio signals. They would continually change the frequency of transmission so that Voyager would always receive the frequency it could accept. Then there should be no problem. But the proof of the pudding ....

On 13 April 1978 they began sending signals to Voyager at the time it should have switched to the backup receiver. Nearly an hour went by — the two-way radio time-lag — and then Voyager answered. The new transmitting technique worked perfectly. So, although partly "deaf", Voyager could continue its voyage of exploration. As a precaution, however, the engineers fed it instructions for automatic encounter procedures at Jupiter and Saturn so that it could send back a limited flow of data even if the backup receiver subsequently failed.

Voyager 2 homes in

Voyager 2's close encounter with Jupiter was scheduled to take place on 9 July 1979. But as early as April, mission controllers were beginning essential pre-encounter operations. For example, they started to reprogram Voyager's computers with new instructions to alter the preplanned sequences of encounter activities. This was done in the light of the remarkable discoveries its sister probe had made.

Among the new activities would be a ten-hour "volcano watch" on Io. Extended studies would be made of the ultraviolet emissions from the plasma torus near Io's orbit, together with extra observations of the dark side of Jupiter after encounter. There would also be a "ring watch", when Voyager passed twice through the plane of Jupiter's equator.

Voyager's observatory phase began on 24 April 1979. For more than a month the imaging system concentrated on taking time-lapse pictures of the Jovian atmosphere. They showed that weather features and the patterns of circulation had changed markedly even in the few months since the previous encounter.

►Voyager 2's close encounter of Europa on 9 July 1979 reveals a network of strange dark markings on the pale icy surface. They appear to be places where the crust has fractured, allowing dark material to well up from below.

▼The surface of Ganymede has dark and light regions, apparently of different age. Closer inspection later reveals that the dark regions are more heavily cratered and presumably older. The light areas have a grooved structure, probably caused by movements of the icy crust.

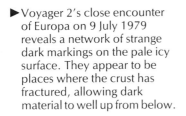

Meeting the bow shock, again and again and again

As Voyager traveled ever closer to Jupiter, the images of the planet it sent back were, if anything, more stunning than those of the earlier mission. On 2 July Voyager encountered the bow shock for the first time, at a distance of about 7 million km (4.5 million miles). Jupiter's magnetosphere had thus expanded considerably since the last encounter. But the magnetosphere proved to be pulsating, because over the next three days Voyager crossed the bow shock no fewer than eleven times!

By 7 July, with Jupiter less than 3 million km (2 million miles) away, improved images were coming in of Callisto, the first of the Jovian moons that would be closely encountered. Another giant ring structure could be identified on it. More distant views of Io showed that three of four volcanoes discovered during the first Voyager encounter were still erupting. The large volcano Pele, however, now appeared to be inactive. And the "hoof-print" — the heart-shaped flow pattern around it — had changed into an oval.

There was hectic activity next day early in the morning PDT as Voyager took the first pictures of Jupiter's ring. Callisto was also closely encountered, and excellent high-resolution

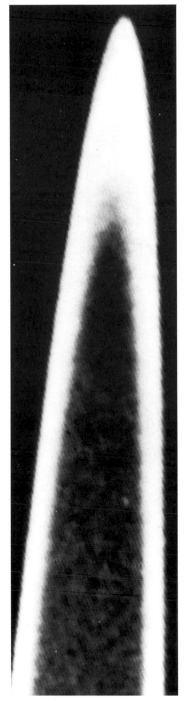

◀Voyager 2, reprogrammed to examine the ring its sister probe discovered, takes this splendid picture. Part of the ring is shown protruding on the right.

▼A later image of the ring shows it to be unexpectedly bright.

pictures were returned. It was displaying a different face from the last encounter, and this face too was heavily pockmarked with craters.

Satellite superlatives

Monday 9 July was a day of encounters, with Ganymede, Europa, Amalthea and Jupiter itself. Ganymede was also presenting a new face, which showed contrasting terrain of youthful, grooved basins and ancient craters. Europa, not clearly seen on the previous encounter, proved to be bright and smooth and covered with dark streaks that looked like cracks, but weren't.

What a fascinating bunch of worlds the Jovian moons proved to be. It seemed that they contained the superlatives among the moons of the solar system: the oldest moon (Callisto), the youngest (Io), the darkest (Amalthea), the reddest (Amalthea and Io), the most active (Io), the least active (Callisto), and now the brightest, whitest and flattest (Europa).

The closest encounter to Jupiter itself came at 3:29 A.M. PDT spacecraft time, when Voyager flew 650,000 km (404,000 miles) above the cloud tops. This was more than twice as far away as its sister spacecraft had been. That far out, Voyager was not expected to experience too much radiation. But, just in case, its computers had been programmed to reset their clocks every hour during encounter. This would prevent the trouble experienced by Voyager 1 (see page 33).

Two hours after encounter, signals were sent to fire Voyager's thrusters for a 76-minute TCM to put the craft in the correct trajectory to encounter Saturn. At much the same time mission controllers lost radio contact with the probe, whose suspect electronics had probably been zapped by radiation. Fortunately, they were able to re-establish contact a few hours later.

Meanwhile, the mission team at JPL were celebrating yet another spectacular flyby. The Jovian system, said Rodney Mills of NASA's Office for Space Science, was a place of "incredible beauty and mystery. Jupiter has been a nice place to go by, but we wouldn't want to stop there — we're going on to Saturn."

And still the images poured in: fine, clear pictures of Jupiter's ring; wispy plumes from volcanoes on Io. Much later, in October, close examination of a Voyager ring picture revealed another tiny moon (1979J1), Jupiter's fourteenth. In 1980, two more tiny moons (1979J2 and J3) were also identified from Voyager pictures. Two orbit closer in to the planet than Amalthea, one orbits just farther out.

How to sum up the remarkable findings made by the Voyagers at Jupiter? Leave it to project scientist Edward Stone. "Just as Galileo's discovery of the large Jupiter moons left a scientific legacy for all of us," he says, "so do the stunning discoveries by Voyager provide a similar legacy for future generations."

A Galilean odyssey

Voyager continued to monitor the Jovian system until 28 August 1979. Not for another 16 years would a spacecraft encounter the giant planet again. That will be the probe Galileo, launched by space shuttle *Atlantis* from Cape Canaveral in October 1989.

In 1995, after traveling to Jupiter by a remarkably circuitous route, Galileo will drop a probe into the atmosphere and go into orbit around the planet for a year or two. What mysteries will this deep-space voyager reveal?

3 | Encounter with Saturn

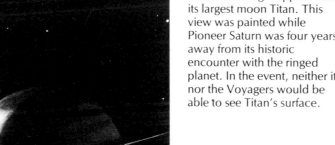

◀ Saturn as it might appear from its largest moon Titan. This view was painted while Pioneer Saturn was four years away from its historic encounter with the ringed planet. In the event, neither it nor the Voyagers would be able to see Titan's surface.

▶ We see different aspects of Saturn's rings as the planet circles the Sun about every 30 years.

Saturn is the most distant of the planets known to ancient astronomers. It lies ten times farther out from the Sun than the Earth. To the naked eye it is nowhere near as striking an object as Venus, Mars or Jupiter. But when spied through a telescope, it surpasses them all. Girdling its equator is a system of beautiful shining rings, whose aspect changes year by year.

When Galileo first trained his telescope on Saturn in 1610, he observed that the planet had "strange appendages" and thought that they must be close-orbiting moons. In 1655 Dutch astronomer Christiaan Huygens suggested that the "appendages" might be a flat ring circling the planet. Twenty years later French astronomer Jean-Dominique Cassini observed a gap in what had hitherto been regarded as a continuous sheet of matter. It became known as the Cassini division.

However, it was soon realised that solid rings would be torn apart by gravitational and centrifugal forces. In the late 1800s scientists found experimental evidence that the rings are actually made up of swarms of particles in largely independent orbits.

Through the telescope
Saturn is the second largest planet after Jupiter, with a diameter nearly ten times that of Earth. Even in a modest-sized telescope, Saturn reveals itself to be unlike the other planets because it displays an elliptical image. A larger instrument separates the rings from the disc and may show other details. It will certainly show that the planet is noticeably flattened at the poles, indicating that, like Jupiter, it is mainly gaseous and rotating rapidly. It has a lower density than any other planet, and would float on water if that were possible.

In many respects Saturn's disc is a pale imitation of Jupiter's. It is crossed by alternate dark and light bands (belts and zones) parallel with the equator. These vary in intensity and color from time to time, but are less well-defined than those on Jupiter.

Astronomers can sometimes observe streaks, filaments and

dark and white spots on Saturn's disc, particularly in the equatorial regions. One of the most conspicuous white spots of modern times was discovered in 1933 by the English comedian Will Hay, who was also an enthusiastic astronomer. Only occasionally do such spots last for more than a few weeks at a time, unlike Jupiter's centuries-old Great Red Spot.

Moons and rings

Dark spots of a different kind reveal Saturn's moons or their shadows in transit across the disc. Through a telescope Saturn appears to have only ten moons — Janus, Mimas, Enceladus, Tethys, Dione, Rhca, Titan, Hyperion, Iapetus and Phoebe, in order of increasing distance from the planet. The biggest moon, Titan, is even bigger than the planet Mercury, while the smallest, Phoebe, is only about 160 km (100 miles) across. Phoebe is unusual in that it has a retrograde motion, circling Saturn in a clockwise direction (from east to west). It is almost certainly a captured asteroid.

Saturn's famous ring system has an overall diameter of some 270,000 km (170,000 miles), or just over twice the planet's diameter. Without its rings, which are highly reflective, the planet would appear very much fainter. We see them from different angles during the 29½ Earth-years it takes the planet to orbit the Sun. This is because Saturn's axis is tilted with respect to the plane of its orbit. The rings are broad but very thin, and when they are presented to us edge-on, they virtually disappear from view.

Three main rings can be distinguished through a telescope. The two outer rings, A and B, are by far the brightest. They are separated by the dark Cassini division. Larger telescopes can make out another small gap near the edge of the A ring. This is called the Encke division after its discoverer, Johann Encke. Inside the B ring is the faint and quite transparent C ring, or crepe ring. There is some evidence of a very diffuse D ring inside the C ring and an E ring beyond the A ring.

Pioneer Saturn: the pathfinder mission

In 1974 the Voyager missions to the giant planets were still in the planning stages. But their pathfinding predecessors, Pioneers 10 and 11, had already blazed a trail at least as far as Jupiter. After Pioneer 11 had encountered Jupiter in December 1974 it was redirected by slingshot gravity-assist on to Saturn.

For such a dual Jupiter-Saturn mission the celestial alignment of the two planets could hardly have been worse. To reach the ringed planet from Jupiter, Pioneer 11 had to travel back on itself across to the other side of the solar

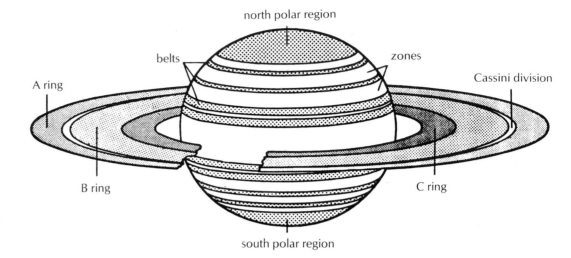

▲ Features of Saturn and its ring system. The disc is very much blander than Jupiter's, but still displays faint belts and ones. The B ring is the broadest, some 25,000 km (15,000 miles) across, compared with 19,000 km (12,000 miles) for the C ring and 15,000 km (9,000 miles) for the A ring.

▼ This curious picture of Saturn is a radio image, which shows that some radiation is associated with the ring structure. Radio signals detected when the Voyager probes began homing in on the planet enabled scientists to determine its period of rotation, 10 hours 12 minutes.

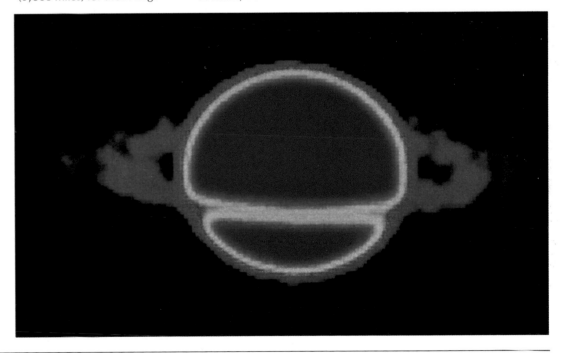

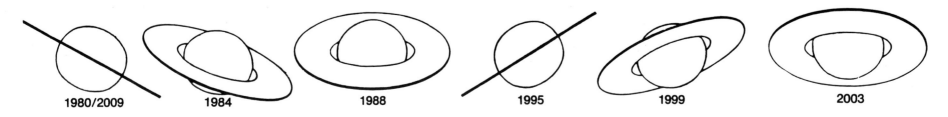

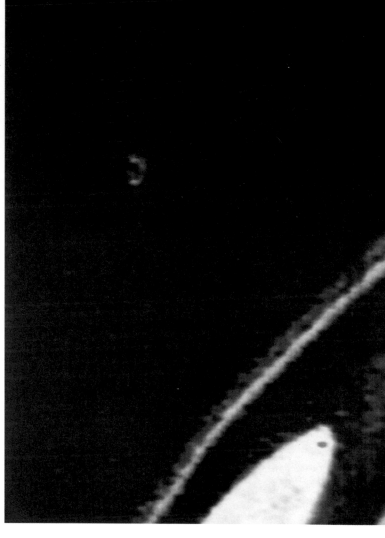

▲ This, the best Pioneer picture of Saturn, is taken on 29 August 1979 when the probe is about 2.5 million km (1.5 million miles) away. It shows the unlit side of the rings, an aspect not visible from the Earth. The A and C rings are visible because they let through scattered sunlight. The B ring, usually the brightest, is dark here because its particles are packed together so densely that they prevent light from getting through.

system. This necessitated a journey that took nearly five years and some very tricky targeting.

In the last few days of August 1979 Pioneer homed in on Saturn and began taking pictures that were 20 times more detailed than any that could be obtained from Earth. To the great disappointment of planetary scientists, however, few details could be seen on Saturn's disc. There was little evidence of the prominent banding there is on Jupiter.

Pioneer made its closest approach to Saturn on 1 September 1979, swooping to within 21,000 km (13,000 miles) of Saturn's cloud tops. A few hours earlier, it had spotted a new ring (F ring) just outside the A ring, together with a new moon (1979S1). It survived two crossings of the particle-strewn ring plane, but came within a cosmic hairsbreadth of colliding with the moon it had just discovered. Mission scientists nicknamed this moon Pioneer Rock.

Among Pioneer's other achievements on this pathfinding mission was the discovery of a second new ring (G ring) far

out between the moons Rhea and Titan. The probe also found that Saturn radiates into space two-and-a-half times more heat than it receives from the Sun. The planet's magnetic field is some 500 times stronger than the Earth's, while its radiation belts are about the same strength.

Mission scientists at Ames Research Center were predictably delighted with the way Pioneer performed throughout this first Saturn encounter. Said Deputy Director Tom Young: "We welcome Saturn into our books of knowledge with a lot of pride that we did it. We can report to Voyager: 'Come on through, the rings are clear!'"

Disappearing rings

By the end of 1979 Pioneer had left Saturn well behind, but was still returning useful data about the solar wind. It would be another 11 months before Voyager 1 would make a close encounter with the ringed planet. During this time Saturn underwent particularly close scrutiny from Earth-bound

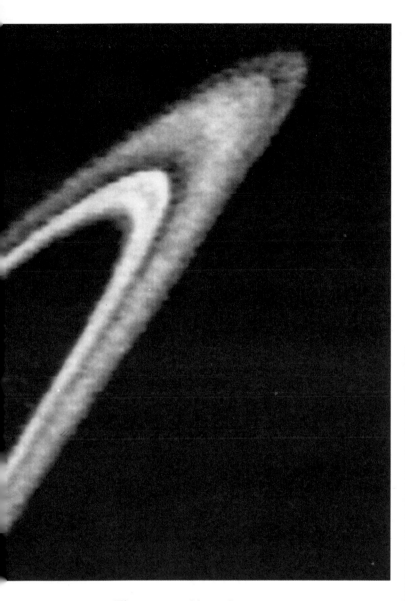

second satellite (1980S6) was discovered in the same orbit as the large moon Dione. These pairs of satellites in the same orbit are termed co-orbitals.

No more new moons were discovered from Earth before the Voyager 1 encounter. The total stood at twelve definite. There were no signs of the disputed Janus; Pioneer had not spotted this planet either.

Voyager 1: the mission continues
Voyager 1 had left Jupiter in March 1979 and had been boosted into a trajectory towards Saturn by the enormous gravity of the giant planet. TCMs in April and October had fine-tuned its path so that it would make close encounters with Saturn and its giant moon Titan on 12 November 1980.

Voyager's instruments switched on to begin a four-month encounter period on 22 August 1980. But not until early October did the planet start revealing its astonishing secrets. At that time nature was staging its own astronomical spectacular in Earth skies. Saturn was appearing just be dawn along with Venus, Jupiter, and the first-magnitude Regulus. It seemed like a good omen.

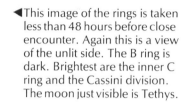

◀This image of the rings is taken less than 48 hours before close encounter. Again this is a view of the unlit side. The B ring is dark. Brightest are the inner C ring and the Cassini division. The moon just visible is Tethys.

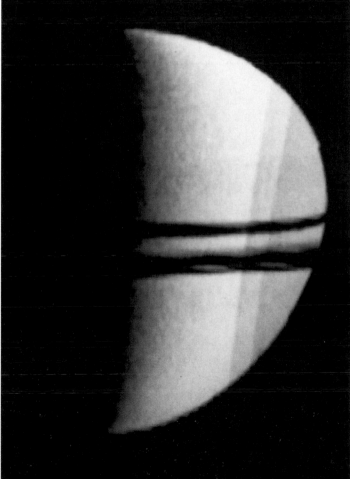

◀Pioneer returns this farewell picture of a crescent Saturn as it begins its journey out of the solar system. Solar activity interfered with Pioneer's signals at this time, resulting in a poorer image than expected.

astronomers. They were taking advantage of an aspect of the planet that occurs about every 15 years, in which the ring system appears edge-on when viewed from Earth.

With the brightness of the rings all but eliminated, astronomers were hoping to discover new faint satellites near the ring edge, like the one Pioneer had discovered. It was fitting that Bradford Smith, head of the Voyager imaging team, and his colleagues at the University of Arizona, should make the first firm discovery early in February 1980. This moon (1980S1) proved to have the same orbit as the one Pioneer had found.

So the two new moons were one and the same. Or were they? In late February astronomers in Hawaii discovered another moon (1980S3). It occupied the same orbit as 1980S1, but was definitely a different object. This was an unprecedented discovery. Here were two satellites in the same orbit on opposite sides of the planet. Within days another example of the same phenomenon came to light. A

On 6 October Voyager was still 50 million km (30 million miles) from Saturn but was returning images that showed a host of new features on the disc and particularly in the rings. Clouds and white and dark spots were becoming visible in the planet's atmosphere. But the rings were beginning to steal the show. They no longer looked like smooth sheets, but were taking on a banded, rippled structure, like a great celestial phonograph disk. Rings began to show up in the dark "gap" of the Cassini division, confirming the Pioneer findings of the previous year. In the B ring dark lanes appeared, extending in a radial direction, rather like the spokes of a wheel. No one had an explanation for this.

Celestial tag

As Voyager continued on its way, greater and greater detail showed up in the rings, revealing that they were made up of a myriad of separate ringlets. Every day Voyager photographed the two co-orbitals (1980S1 and S3), showing that they were actually traveling in slightly different orbits at slightly different speeds — the faster-moving inner one was drawing about 10 meters (33 feet) closer to the outer one every second. What would happen when they met up in about 18 months time?

Mission scientists reckoned that when they came close to each other, gravitational forces would throw the inner moon outward and the outer moon inward. The new inner moon would then be traveling faster than the outer and pull away. About four years later, it would close on the outer and they would once again swop orbits. Such a game of celestial tag must have been going on for billions of years.

Rings and shepherds

On 25 October mission controllers reprogrammed Voyager to devote 10 hours to photographing, once every five minutes, one of the ansae, or ends of the rings. The idea was to produce a movie showing how the mysterious spokes in the B ring behaved. When the images returned were being studied, two tiny new moons (1980S26 and 27) appeared, orbiting on either side of the F ring.

It seemed that in some way these satellites must be responsible for keeping the particles confined into a ring, rather like a shepherd keeps a flock of sheep together. Peter Goldreich, an astronomer at Caltech, had proposed two years earlier that this kind of mechanism might be responsible for Uranus's narrow rings. At Saturn his predicted "shepherd moons" could be seen at work.

By 6 November the world's press had descended on JPL for the encounter and were being treated to one of the most stunning planetary presentations of the space age — animation of the encounter, photographs showing weather in the atmosphere of Saturn, colorful views of Titan, and — the show-stopper — the ring movie showing the radiating spokes. Later that day a 12-minute TCM put Voyager "right on the money" for an accurate close approach.

While Voyager performed like a dream some 1.4 billion km (890 million miles) away, events nearer home were threatening to disrupt mission objectives. The weather

forecast for the Madrid station of the DSN was bad. And on 8 November, four days before closest approach, heavy thunderstorms blotted out six hours of Voyager transmissions. Commented JPL's Ray Heacock wryly at the morning's press briefing: "The rain in Spain has lost all humor for us."

Nevertheless, the imaging data that had been received confirmed the discovery of another new moon (1980S28), just outside the A ring. This appeared to be another "shepherd", keeping the A ring particles in place.

The bottomless pit

On 10 November the weather forecast at the tracking stations for encounter day looked more promising. But mission scientists were strangely subdued, as if readying themselves for the hectic activity that was imminent. There was even an air of frustration among the scientific fraternity, particularly about the dark-spoke phenomenon. Confessed Bradford Smith: "We've never been confused for so long about anything so obvious. It's just so damned frustrating professionally." Referring to the spokes, he asked rhetorically: "How do they form in the first place? How do all of those particles know to turn dark and line themselves up over 25,000 km (15,000 miles)?"

▲ On 18 August 1980 Voyager 1 is still 115 million km (72 million miles) away from Saturn, but is already sending back pictures of exceptional clarity. They show banded features in the northern hemisphere.

◄ An imaginative view of the trajectory followed by the Voyager probes to Saturn. Both craft are in excellent condition when they reach the ringed planet and are ready to repeat the outstanding success of the Jupiter encounters.

►Mysterious dark "spokes" appear in the B ring as Voyager 1 moves in closer. They rotate with the ring, break up and then regenerate in some inexplicable way.

The next day he showed the latest ring pictures and admitted: "The mystery of the ring structure gets deeper and deeper. It seems like a bottomless pit." Clumps were appearing in the F ring and other rings were showing equally eccentric behaviour. Later, another mission scientist remarked that the rings "are not just eccentric; they are stark raving mad!"

Voyager was now imaging many of Saturn's large and small moons, but most attention was focused on the largest, Titan, at the time the only moon in the solar system known to possess an atmosphere. Images showed the atmosphere to be completely covered in dense cloud, with a higher haze layer. In the late afternoon Voyager encountered the bow shock for the first time. This was much farther out than it had been during the Pioneer Saturn flyby. In the evening Voyager swooped in to encounter Titan, passing within 4,000 km (2,500 miles) of it. Disappointingly, the surface remained hidden beneath the clouds. Voyager's infrared and ultraviolet instruments, however, revealed some of Titan's secrets as it probed the atmosphere.

►A scene in the JPL control center of the Deep Space Network during Voyager 1's encounter with Saturn. The latest images are displayed on the monitor screens.

▼It is now 18 October 1980, and Voyager 1's narrow-angle camera can just squeeze in a full image of Saturn and its rings. What a jewel of a planet it is.

Curiouser and curiouser

In the late evening Voyager passed uneventfully for the first time through the plane of the rings. For the next 24 hours it would travel beneath the rings, on the side away from the Sun. It would thus provide a new perspective of the enigmatic rings.

On encounter day, 12 November, mission scientists reported the latest findings to the media at a packed morning press conference. The mysteries continued to deepen as Bradford Smith reported: "In this strange world of Saturn's rings, the bizarre becomes the commonplace, and this is what we saw on the F ring this morning."

Gasps went up from the audience as an arc of the F ring flashed on to the monitor. It was made up not of a single strand, as would be expected, but of two strands intertwined — a phenomenon that came to be called braiding. Smith confessed that the braiding appeared to defy the laws of orbital mechanics, but added: "Obviously the rings are doing the right thing. I guess we just don't understand the rules."

Later, it was Laurence Soderblom's turn to extract gasps from the audience. After showing pictures revealing unusual features on Rhea, Dione and Tethys, he turned to Mimas. This moon is scarred with a huge crater 130 km (80 miles) across — a third of its diameter! Another moon must once have collided with it and nearly smashed it to pieces.

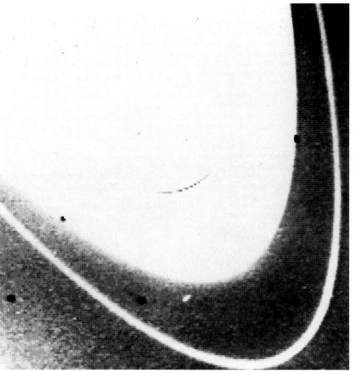

► Saturn's fifteenth moon is spotted in this Voyager 1 picture, taken on 7 November 1980. It is the small white spot, located just 800 km (500 miles) from the edge of the A ring. Very prominent here is the narrow F ring. (The black dots in this and other pictures are reference marks.)

► The icy moon Tethys casts its shadow on the disc of Saturn as new perspectives of the planet unfold. In this picture only the bright A and B rings can be seen, and the disk is clearly visible through the 3,500-km (2,200-mile) wide Cassini division.

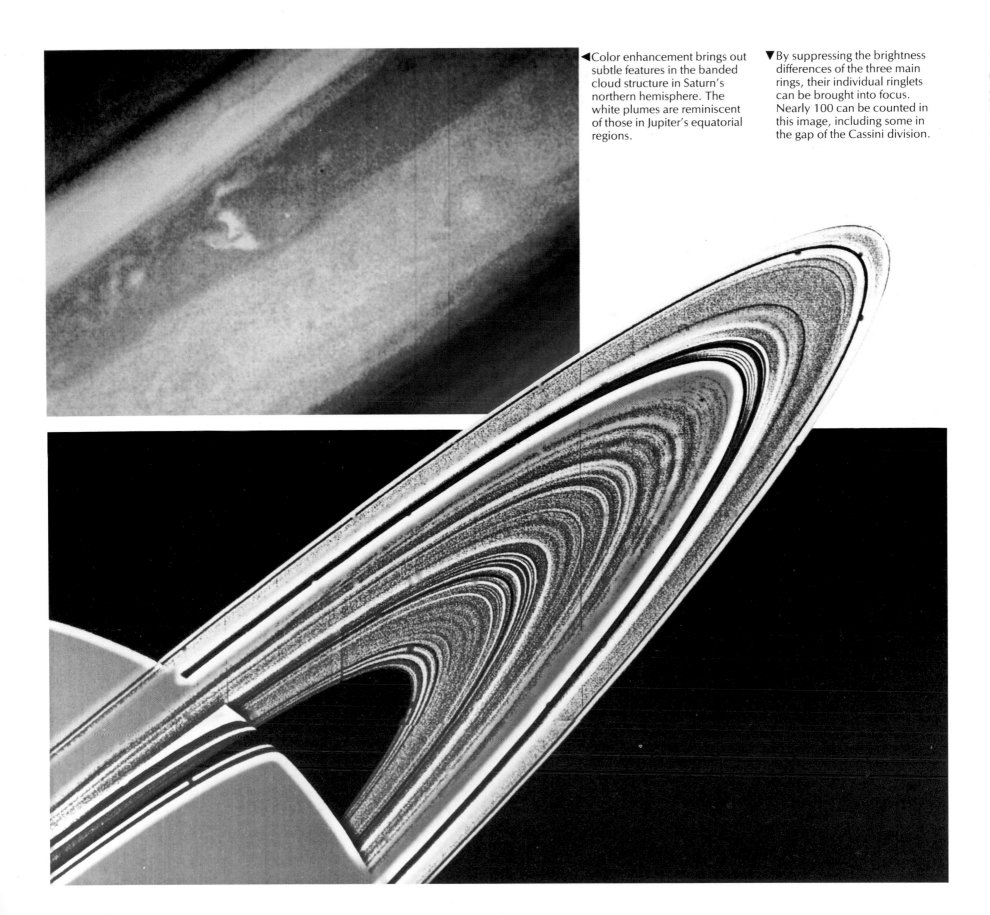

◄ Color enhancement brings out subtle features in the banded cloud structure in Saturn's northern hemisphere. The white plumes are reminiscent of those in Jupiter's equatorial regions.

▼ By suppressing the brightness differences of the three main rings, their individual ringlets can be brought into focus. Nearly 100 can be counted in this image, including some in the gap of the Cassini division.

◀ Just 10 hours remain before closest approach to Saturn, and Voyager is now traveling south of the ring plane and is viewing the unlit side of the rings. This image shows, from the bottom, the B ring, Cassini division, A ring with the Encke division, and F ring. Brightest is the Cassini division, which lets scattered sunlight through. Darkest (shown magenta here) is the dense B ring.

▼ Braiding appears in the F ring, baffling mission scientists. Two narrow bright rings pursue different orbits, twisting around one another like fibers in a yarn. "Knots" also appear, which may be local clumps of material but could be tiny moons.

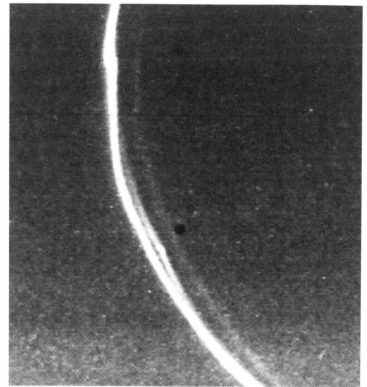

Close encounters of many kinds

In the evening of 12 November Voyager underwent a critical sequence of close encounters in rapid succession, with Saturn, at a distance of 124,000 km (78,000 miles), and with Mimas, Enceladus, Tethys, Dione and Rhea. Voyager's instruments and imaging systems were working at full stretch, as were the mission engineers and scientists who were trying to cope with the streams of data that were flooding back. Many of the results, however, were recorded on tape for playback next day, by which time Voyager had recrossed the ring plane and was once again able to look at the sunlit side of Saturn and its rings.

Among highlights of the findings announced that day was the news that Titan's atmosphere is made up mainly of nitrogen, just like Earth's. The clouds, because of the low temperatures (about −200°C), are probably made up of

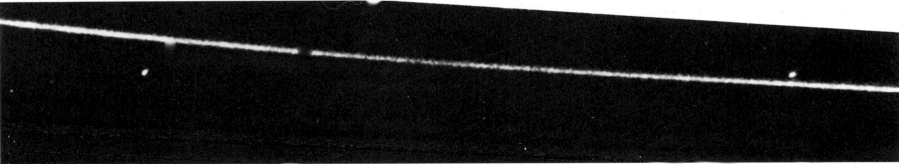

frozen or liquid nitrogen. It was previously thought that the atmosphere would be mainly methane, but this gas is present only in small quantities, rather like water vapor is on Earth. Hydrogen cyanide gas was also detected. This was a significant find, mission scientist Rudolph Hanel explained, because hydrogen cyanide "has long been considered a building block for more complex organic compounds. Let me hasten to add that at these very cold temperatures ... we do not expect any form of life. Nevertheless, we may be dealing with important aspects of organic chemistry."

During the next few days further results and images were released. Among the images, those of Rhea were outstanding. They showed details on the cratered surface as small as 2 km (1.2 miles) across. To achieve such high resolution from the imaging system, the whole spacecraft and the scanning platform were rotated, providing image-motion

compensation (see page 60).

Voyager's Saturn encounter terminated officially on 15 December, leaving mission scientists poring over masses of data and more than 17,500 images. After that the spacecraft began heading out of the solar system. Said chief scientist Edward Stone, appreciatively: "We've had a great ride ... We're in the rush of discovery. Next comes the understanding, which may take years."

Voyager 2 at Saturn
In July 1979 Voyager 2 had left Jupiter behind and begun its two-year cruise toward Saturn. It took much longer to make the journey than its sister craft because it had not approached so close to Jupiter and had not received as great a gravity boost. Voyager's closest approach to Saturn was scheduled for 25 August 1981. Its trajectory was quite different from that

▲(Top) One of the most beautiful of all Voyager pictures, taken the day after closest approach to Saturn. Fine structure can be seen in the rings, and the B ring spokes appear here as bright markings, not dark.

▲(Bottom) The two "shepherd" moons, 1980S26 and S27, seen on either side of the F ring on 16 November as Voyager 1 begins heading out of the solar system. Both moons are only about 200 km (125 miles) across.

of its predecessor. The spacecraft had to be aimed precisely so that it could gain enough energy from Saturn's gravity and be redirected towards Uranus.

It was not expected that this second Voyager flyby would make as many startling disoveries as the first. Rather it would consolidate the data collected earlier. But having seen what the Saturnian system was really like, mission scientists reprogrammed Voyager — for example, to investigate more closely the strange dark spokes in the B ring, which mission scientists now belicved were a phenomenon caused by dust particles being lifted out of the ring plane by electrical forces.

Titan was not a major target for Voyager 2. But Tethys, Enceladus, Hyperion, Iapetus and Phoebe would be looked at more closely. Voyager would also snap some of the smaller moons discovered during the last encounter. It would also target two more new moons (1980S13 and S25) identified from Earth observations in 1980. Both circle in the same orbit as Tethys and are called the Tethys Trojans.

There was every reason to hope that the images Voyager 2 would send back would be better than those from Voyager 1. This was because the Sun had risen higher over the rings, which should thus be much brighter. Voyager 2 would also get much closer to the sunlit side of the planet and the rings.

Stormy weather

The encounter period for Voyager 2 opened on 5 June 1981 and was scheduled to run until 28 September. On the fourth anniversary of its launch, 20 August, the spacecraft was five days and less than 8 million km (5 million miles) away from Saturn. It was hurtling through space at a speed of 11 km (7 miles) a second.

Already the quality of the images being returned was far superior to those taken at a similar stage in the previous encounter. The dark ring spokes were clearly visible, and there was greater activity taking place in Saturn's atmosphere. Oval features thousands of kilometers across showed where massive storms were raging. Later, Voyager confirmed earlier findings that, in jet streams near the equator, wind speeds reached 1,800 km/h (1,100 mph) or more.

Voyager 2's close approaches to Saturn's moons began with Iapetus on 22 August. It proved to be a strange body with a dark leading hemisphere. Two days later came Hyperion, which one mission scientist said resembled "a mouse-gnawed hockey puck".

On encounter day, 25 August, mission controllers calculated that, following a final TCM, Voyager would arrive at closest approach less than 50 km (30 miles) off target, and 3 seconds ahead of schedule. Not bad targeting after a 1.5-billion-km (1-billion-mile) journey! Among results announced that day was the discovery of a huge crater on Tethys more than 400 km (250 miles) across — more than a third of its diameter.

Blinking good

By late afternoon Voyager began one of its most important observations, which would resolve the structure of the rings with unprecedented accuracy. The instrument involved was

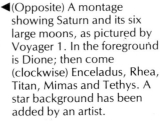

◄(Opposite) A montage showing Saturn and its six large moons, as pictured by Voyager 1. In the foreground is Dione; then come (clockwise) Enceladus, Rhea, Titan, Mimas and Tethys. A star background has been added by an artist.

◄Layers of haze are visible in this false-color picture of Saturn's moon Titan, viewed from a distance of some 22,000 km (13,700 miles). It is the only moon in the solar system with an appreciable atmosphere. With a diameter of 5,150 km (3,200 miles), Titan is the second largest moon in the solar system. Only Jupiter's Ganymede is bigger.

▼Mimas proves to be one of the most heavily cratered bodies we know. Its ancient surface records the battering it received from meteorites and asteroids in the early history of the solar system, some 4 billion years ago. Mimas is about 390 km (240 miles) across.

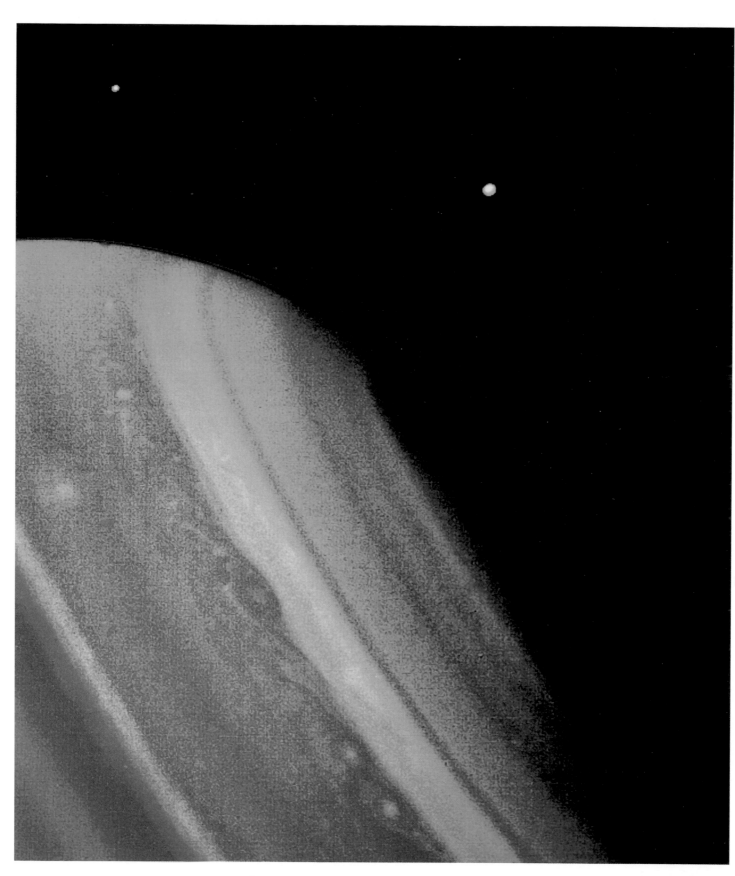

◀Voyager 2 finds more visible signs of activity in Saturn's atmosphere when it heads towards its close encounter in August 1981. On 11 August it images (in false color here) three persistent storm centers in the northern hemisphere. The large, westernmost one is about 3,000 km (1,800 miles) across. The spots are moving relatively slowly, but the ribbon-like jet stream to the north (yellow here) is coursing along at some 530 km/h (330 mph).

▶Two days before closest approach to Saturn, Voyager shoots pictures of the rings through color filters to obtain this spectacular false-color image. The main ring featured here is the diaphanous C ring, with a segment of the B ring at the bottom. The color difference between the two rings is an indication of the difference in the surface composition of the ring particles.

the photopolarimeter, which was aimed at the star Delta Scorpii. This star was so located that it was occulted (covered) by the rings as Voyager sped past. The rings would make the star appear to blink on and off. Such an occultation presented an ideal opportunity for counting the number of rings and estimating the distances between them, and because the occultation would take place in Saturn's shadow, there would be little interference from scattered sunlight.

The occultation experiment lasted for nearly 2½ hours and yielded data on a 82,000-km (55,000-mile) width of ring. The later printout of the ring profile required 800 meters (½-mile) of chart paper! The results confirmed what Voyager's photographs had shown: the rings of Saturn are to be numbered in their thousands, and few regions of the ring system are free of them.

Voyager 2 made its closest approach to Saturn at about 8:25 P.M. PDT, clearing the turbulent cloud tops by 101,000 km (63,000 miles). Then it sped on for a planned ring-plane crossing some 55 minutes later. The chosen crossing point was perilously close to the G ring, missing it by a scant 3,000 km (1,900 miles). In view of the abundance of rings being detected throughout the ring system — gaps and all — mission scientists belatedly wondered whether they had been wise to choose a crossing point so close to the edge of the rings.

They were kept wondering for some time. This was because at 9:00 P.M. Voyager entered Saturn's shadow and its radio

◀ (Opposite) Comparable photographs of Saturn taken by Voyager 2 (top) and Voyager 1 show (in false color) how the planet's appearance has changed over the nine months between encounters. The atmospheric bands are more clearly defined and the rings are brighter in the Voyager 2 image.

▲ The Encke division, the dark gap in the A ring, is not totally free from ringlets. Voyager 2 pictures several in the center of the gap.

▼ Voyager 2 pictures these four tiny and irregularly shaped new moons on encounter day. They are (from the left): 1980S1, one of the co-orbitals); S13, one of the Tethys Trojans; and S26 and S27, the outer and inner shepherd moons of the F ring.

▲ This colorful body is Iapetus, outermost of Saturn's large moons. Its icy crust is covered in part with a dark reddish-brown deposit, which could be some kind of organic material. Iapetus measures 1,440 km (895 miles) across.

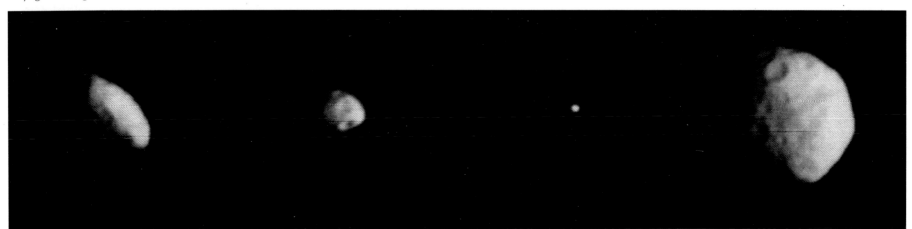

▼Craters and fault lines scar the surface of icy Enceladus, seen here from a distance of 120,000 km (74,000 miles). Enceladus is one of the most reflective bodies in the solar system — more reflective even than fresh snow! It measures about 500 km (300 miles) across.

▶On 29 August 1981 Voyager 2, with scan platform working again, takes this parting shot of Saturn. It shows the unlit side of the rings, with the B ring dark and the Cassini division bright.

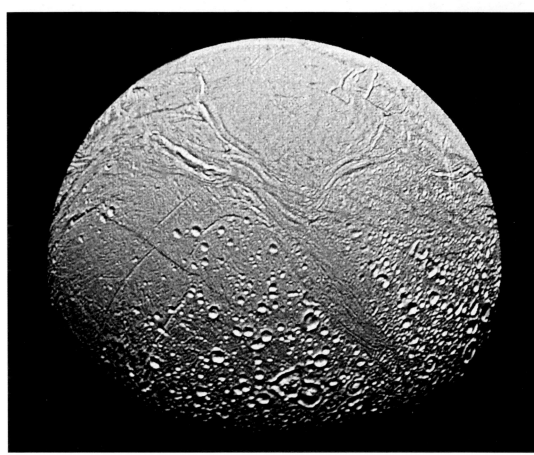

signals were cut off. At JPL, radio blackout came at about 10:26 P.M. local time. Mission controllers would not know until nearly midnight whether the spacecraft had made it through the ring plane. A couple of minutes before midnight, Voyager's faint signals began coming through again, transmitting the data acquired during the blackout period. Mission scientists and engineers were jubilant — Voyager had survived ordeal by ring particles and was setting course for an encounter with the next planet out, Uranus.

Scan jam

The euphoria at mission control was short-lived, however. The telemetry signals coming back from Voyager 2 were far from normal. Engineering data indicated that there had been some spurious firings of the control thrusters during the blackout period. And, what was worse, the movable scan platform carrying the cameras and other instruments had jammed. It could swing up and down but could not slew, or turn sideways. What had happened? Had the electronics failed, or had Voyager been hit by ring particles?

In the early hours of 26 August mission controllers sent signals to turn off the preprogrammed sequencing and put the instruments in a standby mode. Later, they ordered Voyager to play back the stored data to establish when things had gone wrong. First, high-resolution pictures came in of the enigmatic F ring and of one of the F ring "shepherds". But where there should have been high-resolution pictures of Enceladus, there was only blank space. This pinpointed the time of platform-jamming to about half an hour before the ring-plane crossing. A few wide-angle photographs followed that were broadly on target, but after them came nothing but disappointing blank space.

Had ring-particle impacts caused the problem? At an afternoon science meeting, Frederick Scarf presented results from the plasma wave instrument — still working — which had registered a million times more activity than normal while Voyager was crossing the plane of the rings. It was not ordinary plasma radiation, but was almost certainly generated by electrical activity occurring on board the spacecraft itself. Scarf suggested that it could have been caused by the impact of a myriad of microscopic dust grains.

But whatever had caused the anomaly, it had to be rectified if Voyager 2 was to make a successful encounter with Uranus. Mission controllers thought that the best way to tackle the problem was to "exercise" the scan platform little by little, and hope that this would free it. Over the next two days they tried this softly, softly approach, and Voyager responded, if at times jerkily.

On 28 January a happier Richard Laeser opened the daily press conference. "If I have a smile on my face this morning," he said, "it's because there's a good chance that Saturn will be on our TV screens again by the end of the day." In mid-afternoon signals were sent to instruct Voyager's errant scan platform to look back at Saturn again. And right on cue, at a little after 5:30 P.M. local time, a picture appeared on the monitor. It was not quite centered, but it was there. The old team — Voyager and JPL — were back in business.

4 | Encounter with Uranus

On the night of 13 March 1781 a musician and amateur astronomer named William Herschel was observing the stars in the constellation Gemini from his garden in the spa town of Bath, in the west of England. After a while he spotted in his field of view an object that could not be a star because it displayed a definite disc. It was, he wrote in his notebook, "a curious either Nebulous star or perhaps a Comet".

On the following nights he saw the curious object move slowly across the background of stars. Subsequent calculations of its orbit showed that it could not be a comet. What Herschel had discovered was in fact a seventh planet, travelling through the depths of space far beyond the orbit of Saturn, the outermost planet then known. His discovery of the planet, which we now know as Uranus, effectively doubled the size of the known solar system.

In 1787 Herschel spotted the two largest moons of Uranus, Oberon and Titania. He also reported that he had observed a ring around Uranus. We now know that there is a ring system, but Herschel certainly could not have seen it with his equipment. Not until 190 years later, in 1977, was a set of rings spotted, nine thin and very dark ones. They were observed when Uranus passed in front of a bright star whose light flickered as the rings passed by.

In addition to Herschel's Oberon and Titania, three other moons have been discovered — Ariel, Umbriel and Miranda. But virtually nothing can be learned about them through a telescope because they are too far away. Interestingly, Neptune's moons are the only ones in the solar system not named after characters in Greek or Roman mythology. Herschel's son John named the first four moons discovered after characters in English literature. Oberon and Titania are the king and queen of the fairies in Shakespeare's *Midsummer Night's Dream*; Ariel and Umbriel appear in Alexander Pope's *Rape of the Lock*; Ariel also appears in Shakespeare's *The Tempest*, as does Miranda.

▼ Uranus is an almost identical twin of Neptune as far as size is concerned. Both are about four times the diameter of Earth. This artist's impression of the two twin giants was made three years before Voyager 2 encountered Uranus, and shows the planet reasonably accurately, surrounded by its known ring system. The artist had no way of knowing that Neptune would prove to be an altogether more interesting world, with a deep blue color resembling that of Earth.

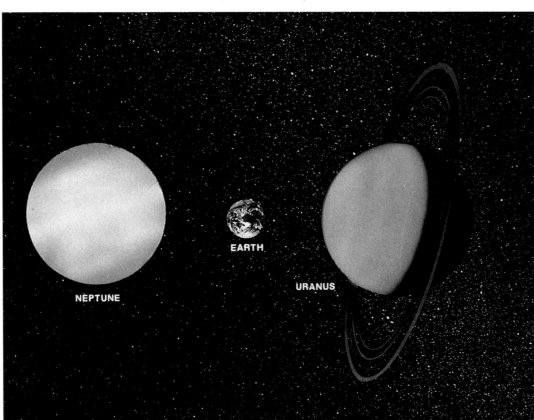

NEPTUNE   EARTH   URANUS

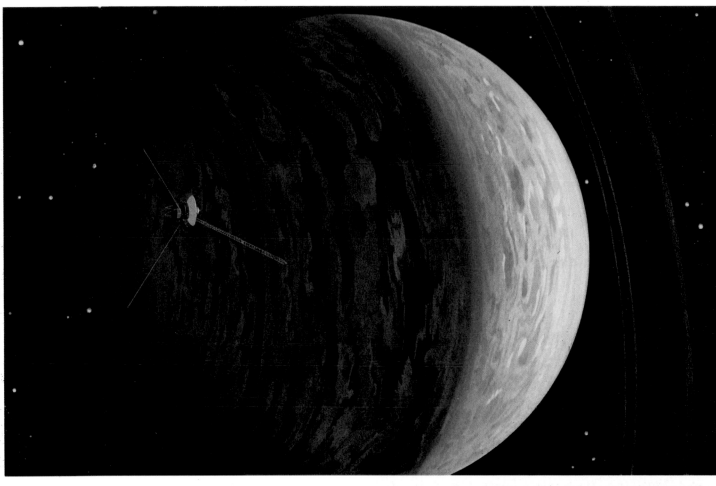

◀This painting of Voyager 2 making its close flyby of Uranus envisaged planet-wide bands of clouds, much like those found at Jupiter and Saturn. In reality the Uranian atmosphere would turn out to be almost uniformly bland.

▼Uranus, its large moons and its rings are pictured in an interesting 3D image. It was obtained in 1984 by processing information recorded by a CCD (charge-coupled device) detector on a telescope at the Las Campanas Observatory in Chile.

Through the telescope

Uranus is the third largest planet, after Jupiter and Saturn, with a diameter about four times that of Earth. But it is so far away — nearly 3 billion km (2 billion miles) — that it appears to us merely as a faint star. It is scarcely visible to the naked eye even when it is at maximum brightness. Through the telescope, Uranus presents a pale bluish-green disc, on which astronomers have occasionally reported vague markings. At such a distance from the Sun, Uranus takes 84 Earth-years to complete one orbit.

Like the other planets, Uranus rotates on its axis, but the orientation of its axis in space is unique. The axis of most planets is tilted only a few degrees from the perpendicular, relative to the plane of their orbit around the Sun. And from Earth they appear to spin almost upright, like a top. The axis of Uranus, however, is tilted at more than a right-angle, no less than 98°! So that from the Earth the planet appears to be rolling on its side, with its poles pointing alternately towards us every 42 years.

In 1986 the south pole was pointing directly towards the Earth. The planet's rings and the orbits of its moons faced us rather like the circles on a bullseye target. And that year the Uranian system was the target for the indefatigable Voyager 2 probe.

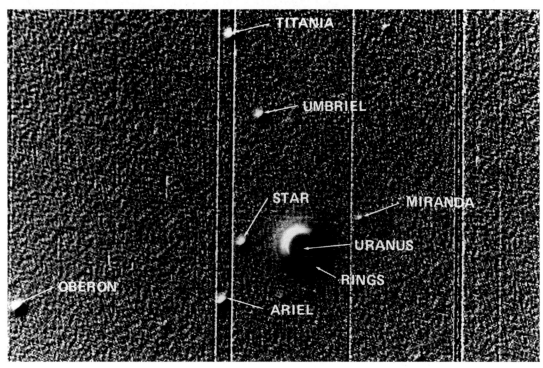

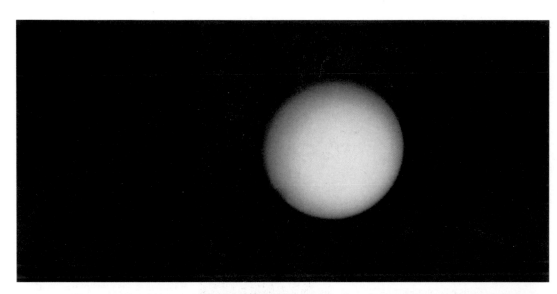

Voyager 2: the mission continues

In August 1981 Voyager 2 passed through the Saturnian system, following in the path of its sister craft Voyager 1. But whereas Voyager 1's encounter days were over, Voyager 2 had other planets to conquer — the mysterious, remote worlds of Uranus and its near-twin Neptune.

JPL engineers carried out painstaking and lengthy analysis and engineering tests to try to establish the cause of the platform jam that had occurred on 25 August, and it took them the best part of three years to be certain of what had happened. They reckoned that the failure had occurred (and could recur) when the platform was slewed rapidly during picture-taking. Under rapid movement, the high-speed gear train on the platform lost its lubricant and seized. After a period of rest, the lubricant migrated back into the gears, allowing movement again.

To prevent the same fault happening during the upcoming encounter with Uranus, the engineers had to rethink their picture-taking strategy. Picture-taking involving slow scans of the platform would not be a problem. But where it would involve rapid scans, they reprogrammed the on-board computers to rotate the whole spacecraft, which would then only necessitate slow platform movement.

Antismear and antiwobble

The technique of rotating the whole spacecraft in this way had already been incorporated into the picture-taking program to facilitate high-speed close-up photography without smear or blur. To take unblurred pictures of an object while traveling past it, the camera has to be moved with the shutter open to follow the object. This technique, called image-motion compensation, was practiced successfully during the Saturn encounter for photographing the moon Rhea. But at Uranus, the problem would be much more acute because the spacecraft would be traveling past the objects it was to photograph at a relative speed of over 65,000 km/h (42,000 mph)! Also, the light levels at Uranus would be only one fourth those at Saturn, because Uranus is twice as far away from the Sun. This meant that the camera shutters would have to remain open for very much longer, increasing the possibility of smear. An additional problem was that the rings and moons of Uranus were known to be dark. Trying to photograph them, said mission scientist Richard Terrile, was "like trying to photograph a piece of charcoal against a black backdrop".

A further measure would be needed to prevent picture-smear. As it flies through space, Voyager wobbles slightly as it reacts to various on-board physical activities, such as the stopping and starting of the tape recorder. To compensate for this wobble, the spacecraft periodically fires its thrusters for 10 milliseconds (10 thousandths of a second) at a time. During long exposure times, these thruster burns could jolt Voyager enough to blur images.

So the on-board computers were reprogrammed to reduce the thruster-burn time to just 4 milliseconds for the encounter period. This would be enough to keep the spacecraft under control, but would significantly reduce image-smearing.

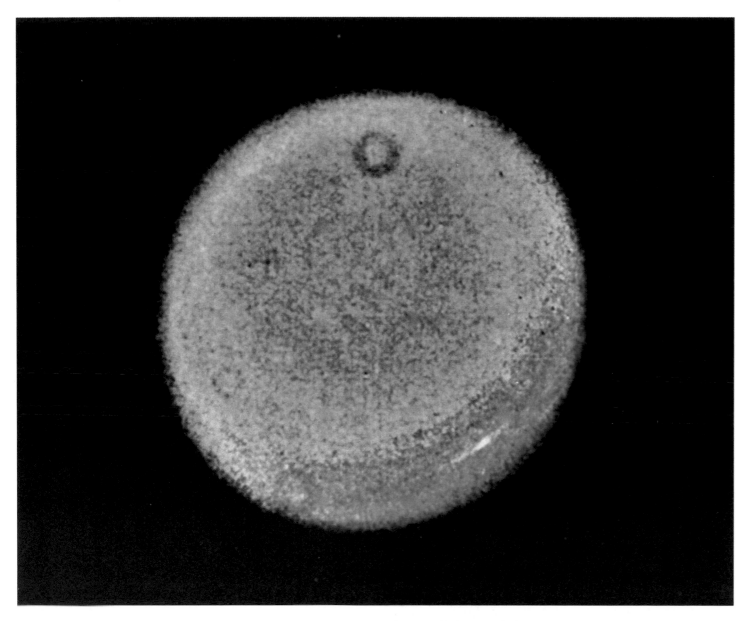

◄(Opposite, top) When still 74 million km (46 million miles), from Uranus, Voyager 2 pictures the planet . There are no signs as yet of any cloud systems in the overall bluish-green atmosphere.

◄(Opposite, bottom) Six images taken by Voyager 2's narrow-angle camera are combined by computer to create the first complete picture of the epsilon ring around Uranus. It is the most prominent of the rings, but is still very dark compared with those of Saturn.

◄Ten days and 13 million km (8 million miles) away from closest approach, Voyager 2 spies a cloud in the atmosphere of Uranus. But it is only visible after extreme computer enhancement. The color differences shown here probably relate to the presence of smog-like particles in the atmosphere. The dark rings in the picture are caused by the presence of dust in the camera optics.

Compressing the data

Images from Voyager are transmitted back to Earth in the form of electronic bits, or binary digits, a sequence of 1s and 0s. But the farther away Voyager is, the lower the quality of the electronic image data received. This is because the data become adulterated with "noise", or interference. But noise can be reduced and image quality enhanced by lowering the rate at which the data are transmitted.

At Jupiter, a transmission rate of over 115,000 bps (bits per second) gave a large number of quality images. But at Uranus, a transmission rate as low as 21,600 bps would be needed to transmit pictures of acceptable quality. At such a slow rate, only a limited number of images could be transmitted per day during encounter because, as originally programmed, each image required over 5 million bits (see page 16).

JPL scientists and engineers overcame this limitation by a clever technique called image data compression. They reprogrammed Voyager's computers to preprocess all imaging data prior to transmission. Instead of transmitting the full eight bits originally required to define the brightness of each pixel (picture-element) of the image, Voyager was instructed to transmit only the *difference* in brightness between each pixel. This technique made possible a 60 per cent reduction in the number of bits required per image, enabling Voyager to send back some 200 quality images per day at Uranus, three times more than it could have done without data compression.

Arraying the antenna

But, whatever the bit rate, the Uranus encounter also presented an unprecedented challenge for radio communications. At the time of the Uranus flyby, Voyager

▼ Two shepherd moons are spotted keeping the particles of the bright epsilon ring in place. The two moons are tiny, only about 20-30 km (12-20 miles) across. Six of the remaining eight rings around the planet are visible, and there are hints of the other two. Later images will reveal a tenth and maybe an eleventh ring.

▶ This picture shows a slice through the epsilon ring, in a region where it is about 30 km (20 miles) across. It is taken on encounter day, 24 January 1986. The structure shows up when the photopolarimeter registers the light filtering through the ringlets from a distant star. The colors are false. The ringlets are in fact very dark.

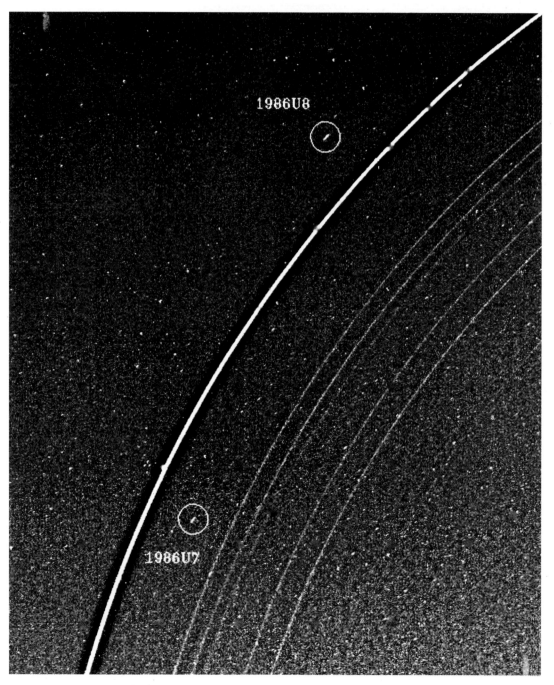

would be nearly 3 billion kilometres (2 billion miles) away from Earth, and its radio signals would require 2¾ hours to make the journey.

For improved reception of the feeble signals reaching Earth, the Deep Space Network of tracking and communications stations had to be updated. New high-efficiency 34-meter (112-foot) antennas were built at Goldstone, California, and at Canberra, Australia.

To optimize reception, all the antennas at each DSN station were be electronically connected. This technique, called arraying, would greatly boost the signal strength. A further boost would be given by linking Australia's 64-meter (210-foot) Parkes radio telescope with the DSN's Canberra array. The Australian connection would be crucial during the encounter, because Voyager would be electronically visible at Canberra and Parkes for long periods.

The encounter begins

The programming of Voyager for data compression, antismear, antiwobble and so on, took place during Voyager's four year interplanetary cruise between Saturn and Uranus. As the spacecraft homed in on its bullseye target, one JPL engineer commented: "We're flying a different spacecraft than we launched." Would the "new" Voyager be able to cope with the challenges topsy-turvy Uranus and its moons would throw at it?

As early as July 1985, with flyby still half a year away, Voyager began to send back images of Uranus better than any obtainable from telescopes on Earth. Not that there was much to see. The planet presented a bland, featureless disc, quite unlike the banded discs of Jupiter and Saturn. Here was quite a different planet.

The far encounter period began on 5 November 1985. All Voyager's instruments were switched on and calibrated. At a leisurely pace, it began to take time-lapse photographs to study any features in the Uranian atmosphere. It began scanning around the planet for its rings and searching for new moons. The first new moon was discovered on 30 December. By 20 January 1986 a further nine tiny new moons had been discovered, bringing the total of Uranian moons to no less than fifteen.

All the new moons circled within the orbit of Miranda, the closest moon to the planet visible from Earth. Seven of them —none larger than 170 km (110 miles) in diameter—proved to be in such similar orbits that they are almost certainly the remains of a single moon that was shattered by a collision many eons ago.

As Voyager approached closer and closer, it took detailed photographs of the nine known rings and discovered traces of a tenth. Two of the new moons were seen to act as "shepherds" for the broadest, brightest and outermost ring, the epsilon. Neither appeared to be more than 20 km (12.5 miles) across.

A crescendo of discovery

Voyager reached a peak of activity in the few days before and after closest approach, searching for and finding a magnetic field, detecting a great atmospheric glow rather like the auroras back on Earth, and photographing in turn at high resolution Uranus's five large moons, Oberon, Titania, Umbriel, Ariel and Miranda.

The closest approach to Uranus came on 24 January 1986. Voyager flew past the planet some 80,000 km (50,000 miles) above the cloud tops. It was 10:00 A.M. PST spacecraft time. Voyager was a little over a minute early. Guiding Voyager to so accurate a rendezvous after a journey of some 5 billion km (3 billion miles) was likened to sinking a golf putt from a distance of 2,500 km (1,500 miles)!

After closest approach, Voyager whipped behind Uranus to view the planet's dark hemisphere. It also performed a detailed investigation of the rings as they occulted (covered) the star Algol, in the constellation Perseus. As streams of data

▼This is the Uranian moon Umbriel, which orbits some 270,000 km (170,000 miles) from the planet. A curious feature shows up on the limb at lower right. It could be a circle of fresh ice around a new impact crater. Umbriel has a diameter of about 1,190 km (745 miles).

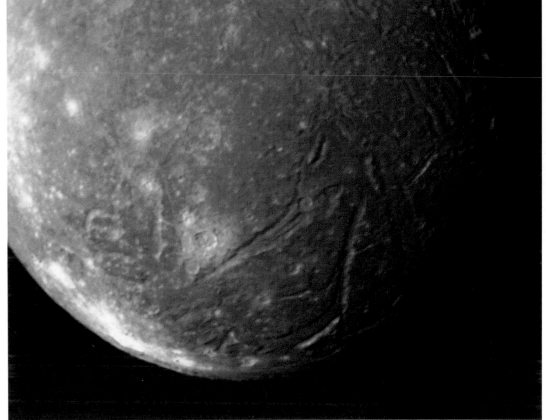

▶This dramatic picture of Ariel reveals great fault valleys and scarps. Some of the younger valleys are filled with brighter material, which presumably has welled up from below. Like the other Uranian moons, Ariel is made up largely of water and methane ice, with a certain amount of silicate rock. It measures about 1,160 km (725 miles) across.

▶If you were orbiting around Miranda in a spacecraft, you would see a view like this. Uranus, over 100,000 km (60,000 miles) away, would dominate the sky. Beneath would be the extraordinary contorted landscape of one of the most remarkable bodies in the solar system.

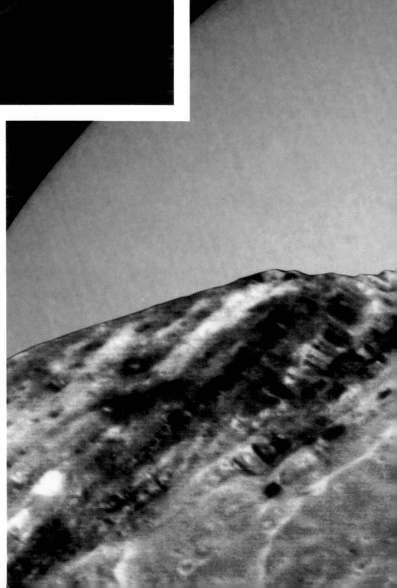

poured in to JPL, project scientist Edward Stone admitted: "We're quite excited. It's the crescendo of discovery."

A letter was read out to the Voyager team from Mrs J. Hole, the Mayor of Bath in England, the home town of the discoverer of Uranus, William Herschel. "Here in Bath," wrote Mayor Hole, "we feel that now, 205 years after the discovery, information will become available that will vastly extend the work commenced by one of the most distinguished astronomers who ever lived."

How right she was. Indeed, Voyager was accumulating much more information than it could transmit in real time, and so stored the excess information on magnetic tape for transmission over the next several days.

Mind-blowing images

Among the imaging data transmitted later were the pictures of Miranda, the smallest and closest of the "big five" Uranian moons. They were, in the words of geologist Harold Masursky, "really mind-blowing". The moon was a "bizarre hybrid", showing certain features of other moons in the solar system. But it had a surface that overall was utterly unique. One area looked like "a stack of pancakes". Another curved almost at right-angles like "a giant race track". Elsewhere, there were deep gorges and steep cliffs.

Measurements of the magnetic field of Uranus showed it to be a little weaker than the Earth's. Also, it was tilted more than 60° to the planet's axis of rotation. Analysis of the radio

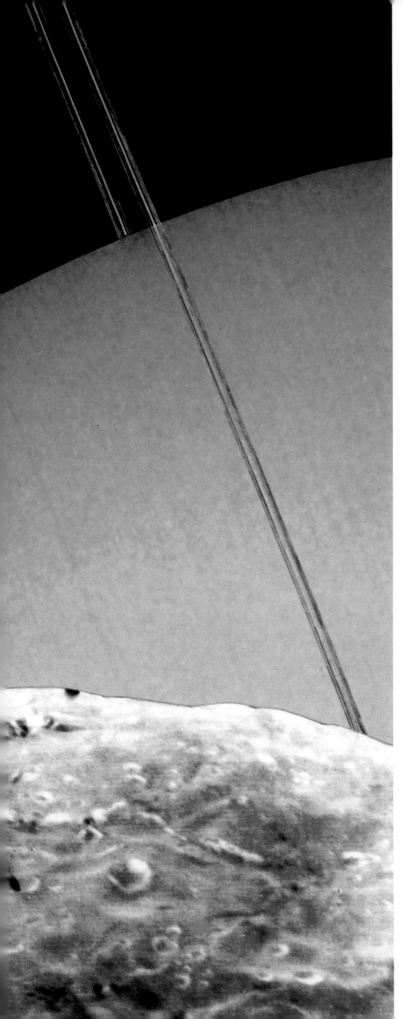

emissions triggered off by the magnetic field allowed the first accurate measurement of the planet's period of rotation (a little over 17 hours).

On 28 January 1986 mission scientists were still poring over the "mind-blowing" pictures of the once mysterious Uranus. But the triumph of one space Voyager was momentarily eclipsed by the tragedy of "seven star voyagers" closer to home. On that day space shuttle *Challenger 2*, with its crew of seven, exploded 73 seconds after lift-off from the Kennedy Space Center. This stopped American manned space flight in its tracks and cast a pall over the whole future of space exploration.

Despite events on Earth, Voyager was outward bound, aiming at a point in space where, celestial mechanics willing, the planet Neptune would be in three-and-a-half years' time. Always assuming, of course, that Voyager 2 would still be functioning.

Voyager mission controller Bruce Bryner had no doubts on this point: "The craft is beautiful. No matter what we throw at it, it keeps on going. You've got to love it!"

▼This is Miranda, smallest and innermost of Uranus's five large moons. Its terrain is utterly unique, with radically different kinds of features abutting one another. Rolling cratered plains suddenly give way to spectacular grooved regions, some oval, some in a chevron pattern. Mission scientists speculate that such a landscape came about following a catastrophic collision with another body. The moon was shattered, and then reformed. Miranda has a diameter of about 485 km (300 miles).

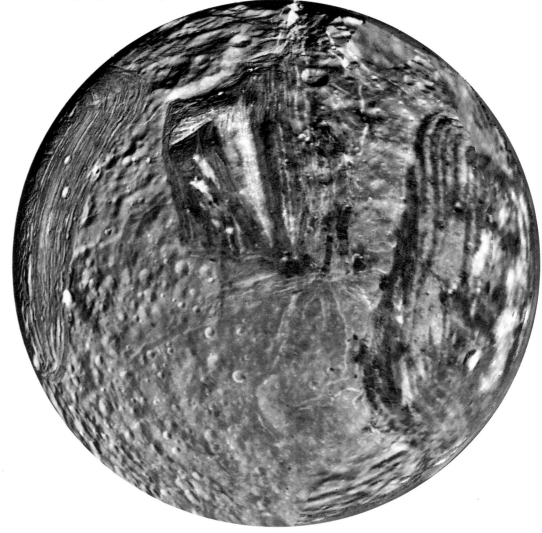

66

Encounter with Neptune

▼ The Parkes radio telescope in New South Wales, Australia, plays a key role in the receipt of signals from Voyager 2 at Neptune. Its antenna measures 64 meters (210 feet) across. It is arrayed with the Deep Space Network's Canberra antenna for greater sensitivity.

In the early nineteenth century astronomers were finding it difficult to work out the precise orbit of the newly discovered planet Uranus. Then it occurred to them that perhaps the planet was being deflected from its expected orbit by the gravitational pull of another planet not yet discovered.

In theory, from the actual orbit of Uranus it should have been possible to calculate where in the heavens the new planet was to be found. But in practice, in those pre-computer days, the calculations needed were formidable.

Nevertheless, two mathematicians, John Couch Adams in Britain and Urbain Leverrier in France, took up the challenge. Independently, Adams in 1845 and Leverrier a year later, they came up with the same result and pinpointed the position of the unseen planet in a particular part of the sky. Surprisingly, both men had difficulty in interesting astronomers in their own countries with their results.

English astronomers eventually began looking for an eighth planet in late August 1846, but with little sense of urgency. Leverrier, meanwhile, decided to write to German astronomers at Berlin Observatory to inform them of his calculations. The Observatory received his letter on 23 September 1846. Immediately after reading it, Observatory astronomer Johann Gottfried Galle became excited, and that very night trained his telescope on the region of the sky Leverrier had indicated. In only about an hour he spotted an unidentified "star". But it was no star — it was the elusive eighth planet, later to be called Neptune.

Through the telescope

In size, Neptune is a near twin of Uranus, being only fractionally smaller. Both are about four times the size of Earth in diameter. Neptune lies nearly 30 times farther away from the Sun than Earth does — so far that it takes nearly 165

Earth-years to complete its orbit.

For most of this time Neptune is the second farthest planet, but at present it is the farthest. This is because since 1979 Pluto has been traveling within Neptune's orbit. And it will continue to do so until 1999.

Even through the most powerful telescope, no details can be seen on Neptune's bluish disc. But we know from instruments that it is made up mainly of hydrogen and helium, with traces of methane. Strangely, it appears to have about the same temperature as Uranus, although it is nearly 1½ billion km (about 1 billion miles) farther away from the Sun.

From Earth we can see only two moons around Neptune — Triton, similar in size to our Moon, and Nereid, just a few hundred kilometers across. Triton has a retrograde motion around the planet in a highly elliptical orbit. Like Nereid, it is probably a captured asteroid. Faint traces of rings have been reported.

Voyager 2: the mission continues

In the last week of January 1986 Voyager 2 passed through the faint ring system of Uranus and, tugged by the planet's gravity, altered course into a trajectory towards its next port of call, Neptune. But the trajectory was not yet quite accurate enough. On 10 February JPL ordered a 2½-hour firing of Voyager's thrusters to accelerate the spacecraft by the few extra kilometers an hour it would need to keep its flyby date with Neptune three-and-a-half years in the future, on 25 August 1989.

During the lengthy cruise period to Neptune, the JPL Voyager team was not idle. As they had done during the cruise period to Uranus, they took the opportunity to reprogram the six on-board computers. They taught Voyager new ways of acquiring, processing and transmitting data.

Their particular concern revolved around the problems posed by the very dim lighting conditions Voyager would find at Neptune, which receives a thousand times less sunlight

◀Astronomer Gerard Kuiper took this classic photograph of Neptune and its two known moons, Nereid and Triton (arrowed), in 1949. The planet itself has had to be overexposed in order to make the moons show up.

◀By late January 1989 Voyager 2 is still over 300 million km (185 million miles) and seven months away from Neptune. But it is sending back images showing what appear to be clouds in the atmosphere.

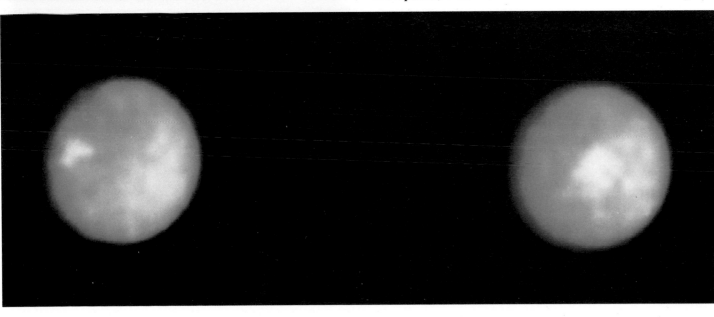

64-meter (210-foot) Parkes radio telescope and the DSN's Canberra antenna, both in Australia, would again play vital roles, because Voyager would be almost directly overhead at these locations during the encounter period. And for the first time the 27 antennas of the Very Large Array radio observatory at Socorro, in New Mexico, would be pressed into space communications service. In all, 38 antennas on four continents would get together to gather the electronic gossamer settling on the Earth from interplanetary space.

A busy year
The reprogramming of Voyager's computers continued throughout the long cruise phase of the mission, as new techniques were tested and refined. Voyager was all the while also returning scientific data about conditions in regions of interplanetary space never before investigated.

In February 1989, with just six months remaining before flyby, a final updating of the on-board programs began. Voyager was being readied for the detailed observation sequences it would perform during the long encounter period.

On 4 May 1989, with just a month to go before Voyager's encounter period was scheduled to begin, JPL received heartening news from the Kennedy Space Center at Cape Canaveral. The space shuttle *Atlantis* had successfully launched their probe called Magellan towards Venus. A few days later a truck with an oversize load left JPL for a four-day journey cross-country to the Cape. On board was JPL's Galileo probe, scheduled for launch in the fall (see page 37). It was going to be a busy year for planetary scientists.

Happy anniversaries
The Voyager Neptune encounter period began on 5 June, when the probe, at 9 A.M. PDT, was about 17 million km (73 million miles) from the planet, and closing on it at the rate of

▲ It is now 17 August 1989, exactly a week before Voyager's flyby of Neptune. The planet is revealing itself as a beautiful object, deep blue and with skeins of white clouds scurrying here and there. But the great find astronomically is the Great Dark Spot, named for its similarity to the Great Red Spot of Jupiter. Note the smaller dark spot farther south.

than we do on Earth. At such low light levels, exposure times for the imaging system would need to be long — up to 15 seconds. And, relative to Neptune, Voyager would be traveling at nearly 100,000 km/h (60,000 mph). To prevent blurring of the image, the antismear and antiwobble techniques practiced at Uranus would again be needed, with additional refinements (see page 60).

More sensitive ears
When it eventually arrived at Neptune, Voyager would be nearly 4.5 billion km (2.8 billion miles) away from Earth. Its radio signals, traveling at the speed of light, would take over 4 hours to reach us. By then the power of its signals would have fallen to a faint one ten-million-million-million-millionth (1/10,000,000,000,000,000,000,000,000th) of a watt. This is many times less than the energy in a falling snowflake!

Several steps were therefore taken to improve reception of the signals. The largest antennas of the Deep Space Network were stripped of their surface and rebuilt 6 meters (20 feet) larger in diameter, to 70 meters (230 feet). As with the Uranus encounter, the antennas would be electronically linked. The

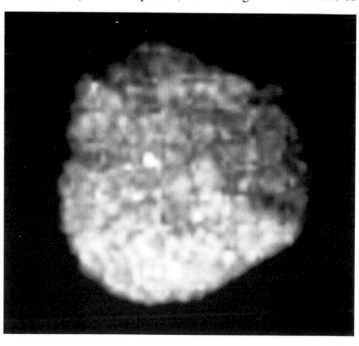

▶Further details on Neptune's disc come into focus as Voyager draws nearer. The Great Dark Spot (GDS) continues to fascinate, its attendant white clouds changing their form day by day. The smaller dark spot near the bottom of the picture, which has a bright core, travels much faster than the GDS within a wind belt speeding along at over 700 km/h (400 mph). The bright feature north of the smaller spot also travels faster than the GDS and is nicknamed Scooter.

◀1989N1, the first new moon Voyager spots, has an irregular shape and is about 400 km (250 miles) across. It proves to be about 30 km (20 miles) larger than the previously known moon Nereid. But the six other moons Voyager finds are tiny, one as small as 55 km (30 miles) across.

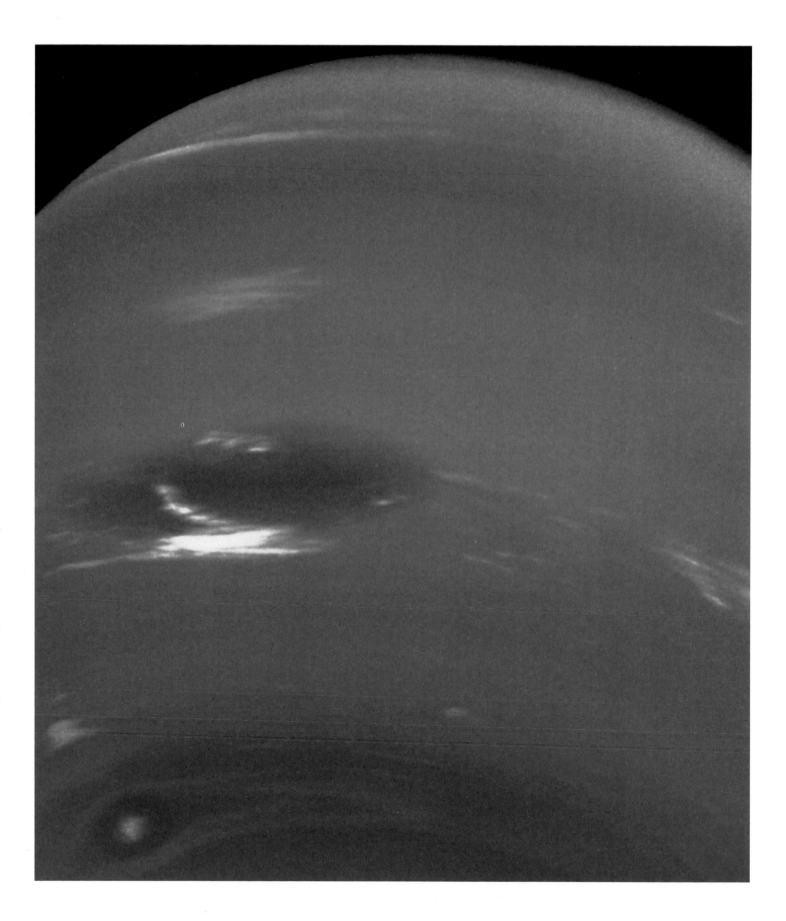

nearly 1.5 million km (900,000 miles) a day. Telemetry from Voyager indicated that all systems were functioning well. On 7 July the first major discovery was announced after processing imaging data received in June. Neptune had a third moon (1989N1).

On 20 July it was the anniversary of the most spectacular of all space achievements, the day in 1969 when astronaut Neil Armstrong from planet Earth took that one small step on the Moon that became a giant leap for mankind. Voyager, the man-made robot, was now making its own giant leap into remote regions of the solar system, where his creators could not, yet, go.

It was now returning increasingly detailed images of the growing disc of Neptune. The disc was not uniformly bland like that of Uranus, but showed several distinct features. These included white spots that could be clouds and a dark oval patch reminiscent of Jupiter's Great Red Spot. It was dubbed the Great Dark Spot. Comparing it with Jupiter's Spot, imaging team leader Bradford Smith said: "It is in about the same proportion to the planet's size and it is at the same latitude."

The presence of so much "weather" on Neptune came as a complete surprise. The winds in the atmosphere reach superhurricane force, traveling at speeds up to 1,400 km/h (900 mph). Strangely they blow mainly from east to west, in the opposite direction to the planet's rotation.

►On 23 August Voyager images complete rings around Neptune for the first time. They include especially bright segments, spotted earlier as ring arcs.

▼On the same day this picture of Triton is returned, with the moon nearly 3 million km (2 million miles) from the camera. It shows interesting, but as yet ill-defined surface features. Its appearance is not unlike that of some of the icy moons of Saturn at the same resolution.

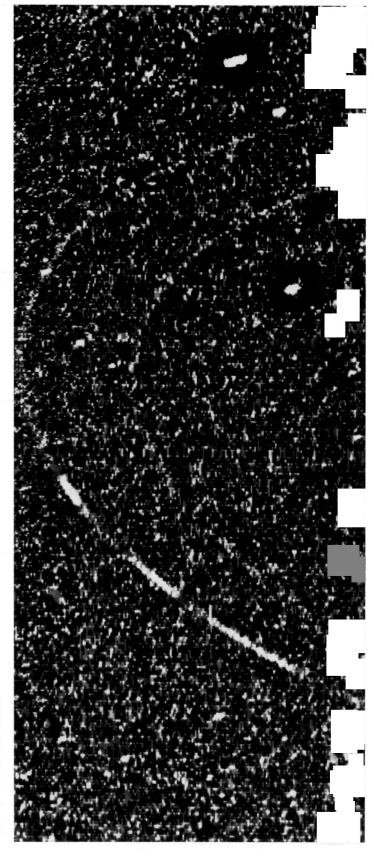

Early in August discoveries started coming in thick and fast. Neptune's largest moon Triton was snapped at a distance, as were three new moons (1989N2, N3 and N4). Soon Voyager was imaging arcs of what appeared to be two incomplete rings around the planet. It was also picking up radio emissions, the first indication that the planet had a magnetic field. From an analysis of the signals, mission scientists determined that the planet rotates once every 16 hours 3 minutes. The rotation period had not been accurately known before.

Sunday 20 August was a very special day for the JPL team — it was Voyager's 12th birthday. Voyager had exceeded its design lifetime by seven years, despite having a dud radio receiver and faulty backup (see page 34). It was still performing with incredible efficiency and bombarding Earth with images of a far-distant world that Earth people will perhaps never see with their own eyes.

The next day, with just four days and 5 million km (3 million miles) to go before the closest approach to Neptune, came a final and crucial TCM. During the past few days, the JPL team had been determining the positions of Neptune and Triton with the greatest accuracy, using images from the on-board camera system. Using these data, they ordered Voyager's thrusters to fine-tune its trajectory so that it would fly as close as possible to the planet and then to Triton.

Still the discoveries came pouring in. Wispy white cirrus

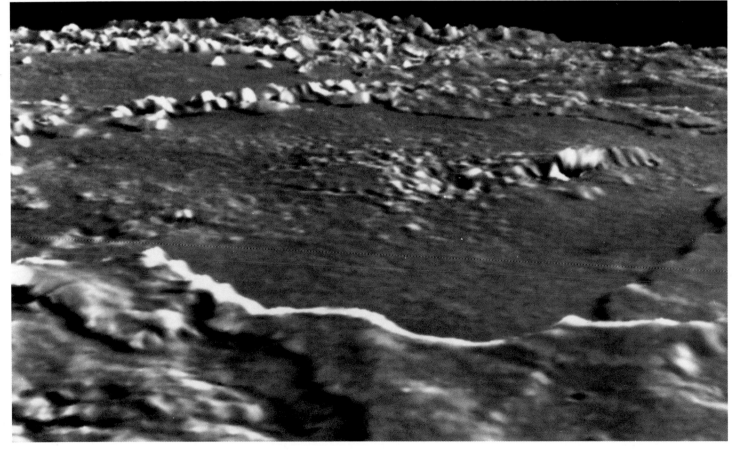

▲ This close-up picture of Neptune's atmosphere, taken just two hours before closest encounter on 24 August, shows high white clouds in relief. Because of the angle of the Sun, they cast shadows on the lower cloud deck. Voyager is now only 157,000 km (98,000 miles) above the cloud tops.

◄ Soon after closest approach, Voyager encounters Triton, which has a truly amazing landscape. This picture shows a filled-in caldera, or crater, from some ancient ice volcano. This is a computer-simulated view, created by exaggerating the topography in the original image, a technique called photoclinometry, or "shape from shading".

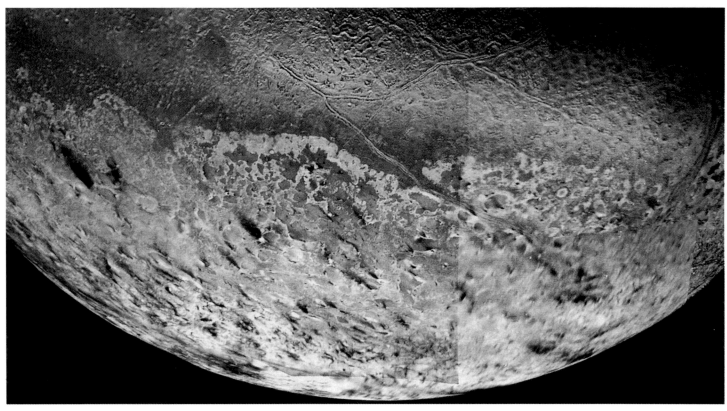

▼This picture is made up of two images taken 1½ hours apart the day after encounter. Voyager is looking back at the Neptunian system and seeing the rings at their best. Two bright rings and a faint inner ring are visible, as is a faint broad band that extends to half-way between the two bright rings.

▲This view of Triton shows the moon's southern hemisphere. It has an extensive polar cap of what appears to be pinkish snow, probably made up of frozen nitrogen and methane. The dark streaks are probably ejecta, material ejected from ice volcanoes. North of the polar cap, the landscape is wrinkled and criss-crossed by long ridges. Triton proves to have a diameter of some 2,720 km (1,700 miles), somewhat less than expected.

clouds scurrying across the equatorial regions were seen to cast shadows on a deeper layer of atmosphere 50 km (30 miles) below. Bradford Smith noted: "This is the first time Voyager has ever been able to see cloud shadows on any of the planets we've looked at." Later, commenting on the mysterious arcs that had appeared in some earlier photographs, Smith said: "The lost arc our imaging-team raiders have been looking for is not an arc. It is Neptune's first complete ring."

A day to remember
Then came the discovery of another two new moons (1989N5 and N6). It was now 24 August, and Voyager was hurtling towards its last planetary rendezvous. Closest approach was now only hours away. NASA Administrator Richard Truly was addressing the Solar System Conference at Caltech that day. Referring to the impending encounter he said: "We, like children, have learned through Voyager's eyes. Surely, this is a day to remember."

A day and a night to remember — for no one at JPL was going to sleep that night. Project scientists, technicians, TV reporters, journalists — all were determined to be in on Voyager's last great picture show.

Among the distinguished VIPs present for the picture show was space scientist James Van Allen. At JPL in the '50s he had designed the instrumentation for the first American satellite, Explorer 1, which discovered the doughnut-shaped regions of radiation around the Earth now called the Van Allen belts. Voyager had entered Neptune's radiation belts earlier in the

day, while more than 600,000 km (375,000 miles) from the planet. The Neptunian magnetic field proved surprisingly weak, weaker even than Earth's. Even more surprising, it is tilted at an angle of 50° to the planet's axis of rotation. If that situation pertained on Earth, our north magnetic pole would be in Los Angeles rather than in northern Canada.

At 9:00 P.M. PDT spacecraft time Voyager dipped down to its closest approach to the cloud-flecked deep blue planet. It sped past just 4,900 km (3,050 miles) above the clouds shrouding the north polar regions. It was by far the closest flyby Voyager had made in its 12 years of exploration. Back on Earth the data from this, Voyager's last planetary encounter, were not received at JPL until just after 1:00 A.M. on the 25th because of the 4 hours 6 minutes time-lag caused by Voyager being 4,398,083,869 km (2,748,802,418 miles) away. Then, deflected by Neptune's gravity, Voyager turned sharply, and streaked on towards Triton, eventually passing within 38,000 km (24,000 miles) of its surface.

Deep-freeze volcanoes

As exciting as the close-up pictures of Neptune had been, they were no match for the breathtaking vistas of Triton that now unfolded on the monitors at JPL. Triton was once thought to be one of the largest moons in the solar system, but Voyager showed it to be smaller although brighter, than expected. Joked Bradford Smith: "Triton has been shrinking as we approached, until we feared that by the time we arrived, it might be gone!"

But Triton was still there, and what a moon it was, with a geology unique in the solar system. The surface is rugged and pockmarked, criss-crossed by cliffs and deep faults, and there is evidence of volcanic craters. But one thing is certain: volcanoes on Triton could not be like those on Earth.

Triton is a deep-frozen world, with surface temperatures as low as −236°C, just 37°C above absolute zero, the lowest possible temperature. This makes Triton the coldest place we know in the solar system, and rules out Earth-type volcanoes pouring out molten lava. Triton's volcanoes would probably spew out liquid and gaseous nitrogen, which would freeze to create the rugged surface terrain we see.

But did the experts think there were Triton volcanoes? Laurence Soderblom reckoned that, judging by the lack of impact craters, the youngest features on Triton could be only a few hundred million years old. "It allows the possibility," he said, "that Triton is active through modern times — it's not necessarily active now, but it could be again in the future."

Grand finale

The images of Triton were some of the most stunning and enigmatic of the many thousands Voyager had sent back during its 7-billion-km (4.4-billion-mile) space odyssey.

Perhaps we should have expected Voyager to stage something spectacular like this for its final appearance. It obviously intended to go out "with a bang, not a whimper".

Speaking for the imaging team, as Voyager set course for the stars, Soderblom exclaimed: "All we can say is: 'Wow! What a way to leave the solar system!'"

▼ Voyager takes this dramatic farewell picture of a crescent Neptune and Triton three days after close encounter. Soon its cameras will be switched off, and it will begin a journey into deep space that will last an eternity.

Off to the stars

When in August 1989 Voyager project scientist Edward Stone closed the final press conference at JPL at the end of Voyager 2's final encounter, with Neptune, he quoted these lines from T.S. Eliot:

Not fare well,
But fare forward, voyagers.

The days of planetary encounters are over for the aging Voyager probes, but their work is by no means done. They are still in remarkably good condition and have enough fuel and power to last them probably until the year 2020. We should be able to remain in touch with them until then.

Their present and future investigations are organized under what is called the Voyager interstellar mission (VIM). It makes use of many of the same facilities as before, but operates with much reduced staffing. Among the objectives of the VIM is a continuous study of the fields, particles and waves in the distant reaches of the solar system. The spacecraft will also use their ultraviolet instruments to study certain stars.

Sometime within the next decade the Voyagers are expected to detect a sudden change in the space environment, in an area known as the termination shock. It marks the beginning of a region called the heliopause, in which the solar wind first encounters the interstellar environment. Between 10 and 20 years later, the Voyagers will cross the heliopause and pass into interstellar space. Mission scientists are hoping desperately that the instruments and radios will still be operating then so that they can learn what space is like between the stars.

Encounters of the stellar kind
The Voyagers are destined to travel through interstellar space for an eternity, wandering through the stars of the galaxy, we know as the Milky Way.

They are both headed in a different direction. Voyager 1 is moving towards the northern star constellation of Camelopardalis, the Giraffe. In about 40,000 years it will make a far (1.6 light-years) encounter with a star called AC+79 3888. At much the same time Voyager 2 will pass within a similar distance of the star Ross 248, heading towards the constellation Canis Major, the Great Dog. In about 300,000 years time it will pass about 4 light-years away from the Dog star, Sirius, the brightest star we can see in the heavens.

►Neptune, final port of call for Voyager 2. Sending back splendid images to the last (this one is in false color), the indefatigable robot has made its Grand Tour, the like of which will not again be possible for a century and a half.

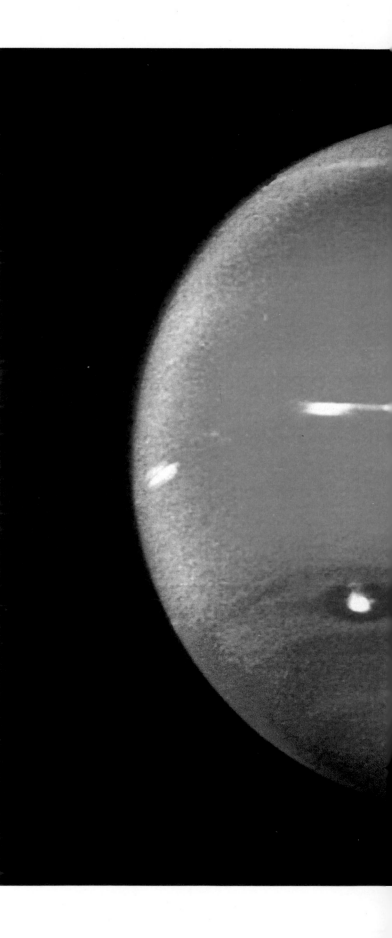

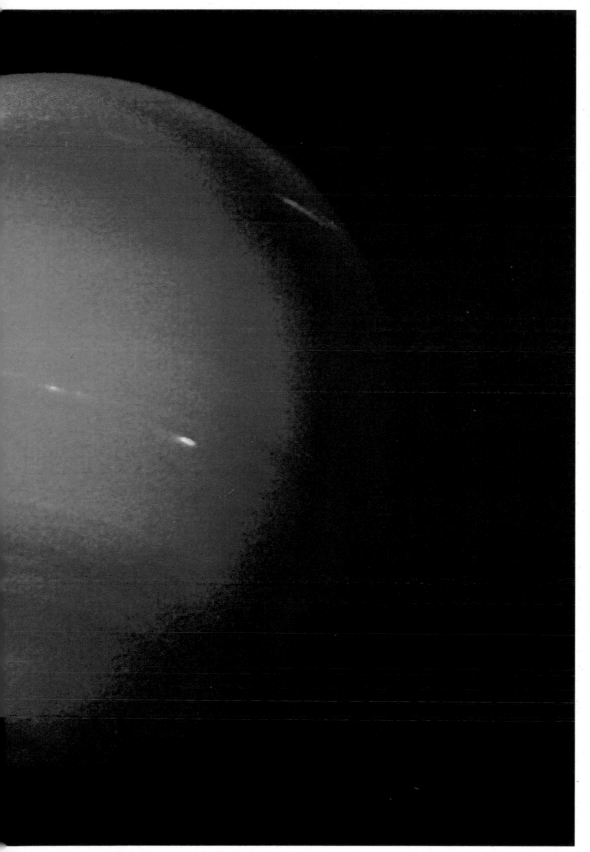

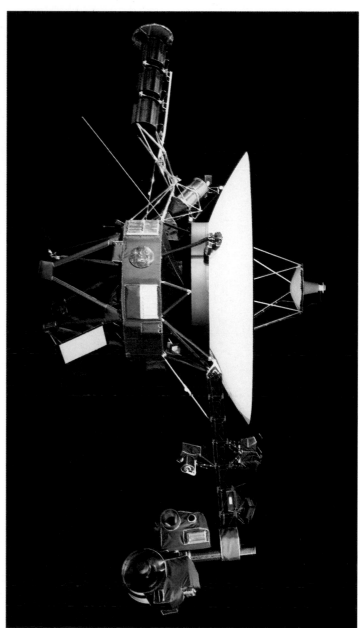

▶ Helpfully, detailed instructions for playing the record are given in ingenious pictorials on the record cover. The "star burst" feature pinpoints the location of our solar system within our galaxy by reference to numerous quasars, the beacons of deep space.

▲ The Voyager record, carrying into the cosmos the sights and sounds of a planet named Earth that circles a star named the Sun. A product of one technologically advanced civilization, it is ready to be played by another, perhaps eons hence, perhaps never.

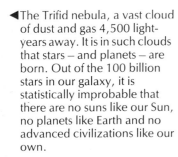

◀ The Trifid nebula, a vast cloud of dust and gas 4,500 light-years away. It is in such clouds that stars – and planets – are born. Out of the 100 billion stars in our galaxy, it is statistically improbable that there are no suns like our Sun, no planets like Earth and no advanced civilizations like our own.

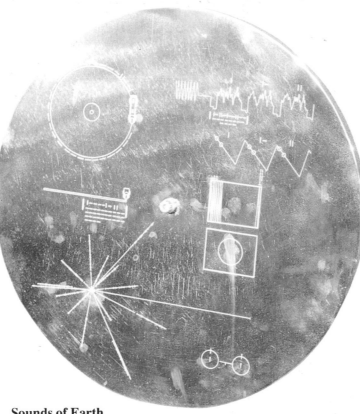

Sounds of Earth

It is just possible, although most unlikely, that one of the Voyagers may in the remote future venture close to another planet in another solar system, where it may be found by another intelligent life form. The alien beings that find it will be bursting to find out where it came from and who sent it. So, for their edification, the Voyagers carry a specially produced phonograph record that encapsulates the myriad aspects of life on Earth.

The 30-cm (12-inch) record is called "Sounds of Earth". It contains greetings from Earth people in many different languages, samples of music from different cultures and eras, and sounds from the human world and from nature — laughter and heartbeats, crying and kissing, wind and thunder, trains whistling and whales singing. The record also contains electronic information that an advanced technological civilization could convert into diagrams, pictures and printed words.

The record was the brainchild of the distinguished astronomer Carl Sagan of Cornell University, Ithaca, New York. The selection of material on it was made by an advisory committee that included prominent scientists and musicians. Sagan was also responsible for the plaques aboard the Pioneer 10 and 11 spacecraft, which carried a message for extraterrestrials in pictorial form.

"Because space is very empty," said Sagan, "there is essentially no chance that Voyager will enter the planetary system of another star. The spacecraft will be encountered and the record played only if there are advanced spacefaring civilizations in interstellar space. But ... the launching of this bottle into the cosmic ocean says something very hopeful about life on this planet."

A phonograph record was chosen as the preferred medium for disseminating terrestrial information for extraterrestrials because it can carry a great deal of data in a small space. In addition, the year of the Voyager launches, 1977, was the 100th anniversary of the invention of the phonograph by American superinventor Thomas Alva Edison.

Each record is made of gold-plated copper and has an aluminum protective jacket, which carries instructions in scientific language on how the record is to be played, using the cartridge and needle provided. Playing speed is 16⅔ revolutions per minute, and the record runs for nearly two hours. It was prepared for NASA as a public service by Columbia Records.

The record begins with 115 photographs and diagrams in analog form. They were chosen to convey some hint of the richness of our civilization. They feature aspects of mathematics, chemistry, geology, biology, technology and human society, and include details of the solar system, descriptions of DNA and human chromosomes, photographs of Earth, the Voyager launch vehicle and human beings of many countries in various settings and endeavors.

This is followed by spoken greetings in some 60 languages and a sound essay on the evolution of the planet Earth, including sounds of weather and surf, Earth before life evolved, life before Man, and finally the development of human civilization. Music follows next, with a selection that reflects the cultural diversity of Earth, of many times and many places. It includes Eastern and Western classical music, ethnic music, jazz, and rock and roll.

Casting into the cosmos

On 16 June 1977 American President Jimmy Carter penned these words at the White House for inclusion on the "Sounds of Earth" record. It is a beautiful, hopeful message that extends a hand of friendship across the stellar frontier, across spans of time measured in eons and distances measured in light-years. Should some other beings, somewhere, somewhen, ever read it, that would be the triumphant final legacy of Voyager.

This Voyager spacecraft was constructed by the United States of America. We are a community of 240 million human beings among the more than 4 billion who inhabit the planet Earth. We human beings are still divided into nation states, but these states are rapidly becoming a single global civilization.

We cast this message into the cosmos. It is likely to survive a billion years into our future, when our civilization is profoundly altered and the surface of the Earth may be vastly changed. Of the 200 billion stars in the Milky Way galaxy, some — perhaps many — may have inhabited planets and spacefaring civilizations. If one such civilization intercepts Voyager and can understand these recorded contents, here is our message.

"This is a present from a small distant world, a token of our sounds, our science, our images, our music, our thoughts and our feelings. We are attempting to survive our time so we may live into yours. We hope someday, having solved the problems we face, to join a community of galactic civilizations. This record represents our hope and our determination, and our good will in a vast and awesome universe."

Glossary

acquisition
Making contact with a spacecraft so that signals can be transmitted to it or received from it.

aimpoint
The point in space that a space probe is aimed at so as to make its intended planetary or satellite encounter.

asteroids
Also called minor planets and planetoids; small rocky bodies that circle the Sun in a broad "belt" between the orbits of Mars and Jupiter.

astronomical unit (AU)
The distance between the Earth and the Sun; approximately 150 million km (93 million miles).

atmosphere
The layer of gases surrounding a planet or a satellite.

attitude
The position of a craft in relation to something else, for example, the horizon.

aurora
An electrical display of colored lights and streamers, seen on Earth mainly in the polar regions as the Northern Lights or the Southern Lights. Auroras have also been detected on other planets.

backup
An item of equipment that can take over the job of another if it fails.

bow shock
The boundary region around a planet where the solar wind meets the planet's magnetic field.

burn
The firing of a rocket motor or thruster.

celestial
Relating to the heavens.

cosmic rays
Highly energetic charged particles that come from outer space.

cosmos
An alternative name for the universe; space.

eccentric
Noncircular; elliptical.

DSN
NASA's Deep Space Network of tracking stations, used for tracking and communicating with distant space probes; operated by JPL.

encounter
The meeting in space between a space probe and its target.

escape velocity
The speed a spacecraft must reach to escape completely from the Earth, or another body.

extraterrestrial
Not of Earth; a being from another world.

flyby
A space mission in which a probe flies past a planet or a moon without landing.

gravity-assist
A method that uses the gravitational attraction of one planet to increase the speed of a spacecraft so as to take it on to another planet in a shorter time than would otherwise be possible.

inner planets
The planets closest to the Sun — Mercury, Venus, Earth and Mars; also called the terrestrial planets.

interplanetary
Between the planets.

interstellar
Between the stars.

ion
An atom or group that has lost or gained electrons and thus acquired an electric charge.

JPL
Jet Propulsion Laboratory, located at Pasadena, California; home of the Voyager project.

launch window
The period of time during which a spacecraft can be launched so as to reach its intended target.

light-year
The distance light travels in a year, nearly 10 million million kilometers (6 million million miles); used as a unit of measurement in astronomy.

magnetopause
The region where a planet's magnetosphere interacts with the solar wind.

magnetosphere
The huge "bubble" around a planet where its magnetic influence can be felt.

meteoroids
Tiny particles of rocky matter that stream through space.

mission
A space flight.

NASA
National Aeronautics and Space Administration.

nominal
A term that means that everything is as it should be.

orbit
The path of one body traveling around another in space.

outer planets
The planets Jupiter, Saturn, Uranus, Neptune and Pluto.

plasma
A gas cloud in which the gas exists as a mixture of ions and electrons.

probe
A spacecraft that escapes from Earth and goes to explore distant planets and moons.

propellant
A substance burned in a rocket engine to propel it.

radiation belts
Intense regions of radiation around planets that have a magnetic field, caused by the presence of highly energetic charged particles.

radio telescope
A telescope, usually with a large metal dish, designed to gather radio waves from the heavens.

redundancy
The duplication of vital parts in a system, so that if one part fails, another can take its place.

resolution
The ability to distinguish details in, say, a photograph or image.

retrograde
Backward; in the opposite direction from normal.

RTG
Radioisotope thermoelectric generator, the source of electrical power on deep-space probes.

satellite
A small body that orbits around a larger one in space; a moon.

scan platform
The movable part of the Voyager spacecraft, which carries instruments that need to be accurately pointed.

"shepherd" moon
One that orbits near a planetary ring and helps keep the ring particles in check.

solar
Relating to the Sun.

solar system
The family of the Sun, which includes the planets, their moons, asteroids and comets.

solar wind
The stream of charged particles continually being given off by the Sun.

speed of light
Approximately 300,000 km (186,000 miles) per second.

stellar
Relating to the stars.

TCM
Trajectory correction maneuver.

telemetry
The transmission of instrument readings and other data between a spacecraft and ground control.

terrestrial
Relating to Earth.

thruster
A small rocket motor used on spacecraft for maneuvering.

time-lag
The finite time it takes for radio signals to travel between a spacecraft and Earth.

torus
A solid geometrical figure with the shape of a ring doughnut.

tracking
Following the path of a spacecraft through space.

trajectory
The path of a spacecraft through space.

trajectory correction maneuver (TCM)
Firing thruster rockets on a spacecraft so as to change its speed and therefore its trajectory through space.

Voyager chronology

1962 In December NASA's Mariner 2 spacecraft becomes the first successful interplanetary probe, when it reports on the hellish conditions on Venus.

1965 Caltech graduate student Gary Flandro puts forward the idea of gravity-assist to reduce journey times to distant planets.

1969 The United States Congress approves the Pioneer project to send two reconnaissance craft through the asteroid belt to encounter Jupiter. The project is assigned to Ames Research Center. NASA also begins to plan a Grand Tour of the outer planets.

1971 NASA selects teams of scientists to support a number of Grand Tour missions to all the outer planets, but by the year's end, the original plans have to be drastically curtailed for budgetary reasons.

1972 Congress approves the revised plans, for two launches to Jupiter and Saturn, project name Mariner Jupiter Saturn. The MJS project begins officially on 1 July. Meanwhile, the pathfinding Pioneer 10, launched on 2 March, is heading towards the asteroid belt.

1973 On 5 April Pioneer 11 lifts off to follow its sister craft to Jupiter, but in an orbit that could allow it to continue to Saturn. On 3 December Pioneer 10 makes its closest approach to Jupiter.

1974 Pioneer 11 makes its closest approach to Jupiter on 2 December, then successfully enters an orbit that will take it to Saturn.

1977 On 20 August Voyager 2 begins its journey into deep space. Voyager follows on 5 September, in a faster trajectory.

1978 On 23 February Voyager 1's scan platform jams, putting the mission in jeopardy, but by May the problem has been rectified. On 5 April Voyager 2's primary radio receiver fails, as does circuitry in the backup. But by modifying signal transmission techniques, full communications are restored on 13 April.

1979 On 5 March Voyager 1 flies within 278,000 km (173,000 miles) of Jupiter. One-way radio time-lag: 47 minutes. Voyager 2 makes its closest approach to Jupiter on 9 July, passing 650,000 km (404,000 miles) over the cloud tops. Pioneer 11, farther out, keeps its rendezvous with Saturn on 1 September. It swoops within 21,000 km (13,000 miles) of the atmosphere.

1980 Voyager 1 makes its close encounter with Saturn on 12 November, at a distance of 124,000 km (78,000 miles). One-way radio time-lag at Saturn: 1 hour 26 minutes.

1981 On 25 August Voyager 2 flies within 101,000 km (63,000 miles) of Saturn.

1983 Pioneer 10 crosses the orbit of Neptune, at present the outermost planet, and in theory becomes the first spacecraft to leave the solar system.

1986 On 24 January Voyager 2 encounters Uranus, flying some 80,000 km (50,000 miles) above the cloud tops. One-way radio time-lag: 2 hours 45 minutes.

1989 Voyager 2 makes its last planetary encounter, with Neptune, on 24 August. It skims just 4,900 km (3,050 miles) above the deep blue atmosphere. One-way radio time-lag: 4 hours 6 minutes.

Index

Acknowledgements/ Credits

The author would like to extend his grateful thanks to his many friends at NASA Headquarters, Washington DC; the Kennedy Space Center, Cape Canaveral; the Johnson Space Center, Houston; and the Jet Propulsion Laboratory, Pasadena; for their unstinting help, advice and encouragement in the preparation of this book.

Almost all of the photographs illustrating this book have been provided by NASA, to whom many thanks. Picture research was by Spacecharts.